Basic
Radiation
Biology

Basic
Radiation
Biology

SECOND EDITION

DONALD J. PIZZARELLO, Ph.D.

Professor in Radiology,
New York University Medical Center,
New York, New York

RICHARD L. WITCOFSKI, Ph.D.

Professor in Radiology, Associate in Neurology,
The Bowman Gray School of Medicine of Wake Forest University,
Winston-Salem, North Carolina

LEA & FEBIGER · 1975 · PHILADELPHIA

Library of Congress Cataloging in Publication Data

Pizzarello, Donald J.
  Basic radiation biology

  Includes bibliographies and index.
  1. Radiobiology. I. Witcofski, Richard L., joint
author. II. Title.
QH652.P59 1975   574.1'915   75-9756
ISBN 0-8121-0522-2

PRINTED IN THE UNITED STATES OF AMERICA

Published in Great Britain by Henry Kimpton Publishers, London

To our friend, *Dr. Isadore Meschan*

A book is the product of many minds apart from those of its authors. Not the least among them are those of dedicated teachers in all fields, who in critical, though often subtle ways help train and shape authors' minds and give them the sturdiness of purpose and depth of motivation needed to complete so lengthy an undertaking. We want to dedicate this edition to just such people, some, like those named below, we can remember, but equally to those whose names, though not their influence, have faded from memory.

ELDA ANDERSON

CECIL EWING

MYRON FAIR

IRENE . GAMBINA

GRACE MacDOUGALL

LOUIS S. MARKS

PAUL MEMMO

SHIRLEY E. ROSSER

ELIZABETH SANTEN

NORMAN SULKIN

ALEXANDER WOLSKY

PREFACE

As in our first edition, we intend this book as a teaching tool for an *introductory* course in radiation biology. Among those who should find it useful are college undergraduates, beginning graduate students, medical students, residents in radiology, radiologists, and radiation physicists who have interest in radiation biology.

We have continued to use a narrative as opposed to a mathematical approach to material where either would do. We have kept the illustrations simple and, hopefully, have improved over those that appeared in our first edition. We have taken a conceptual approach, discussing general principles and important techniques as they apply broadly, rather than presenting masses of specific detail. In accordance with that outlook, we have tried to cover completely the field of radiation biology as it applies to cells and multicellular levels of organization (excluding plants), but we do not intend or claim this work to be an exhaustive treatment of the field as a whole or even of any of its parts.

Many changes have been made from the first edition. This book, unlike its predecessor, does not divide the subject into physics and biology but tries to stress the unity of these in radiation biology. In the first edition there were short reviews of the physics and biology underpinning radiation biology, and much of this has been dropped in this edition. Today's student generally is better informed in these areas than has been true of students in the past, and such a review seems less useful now.

Much new material has been added, and much existing material has been updated. Techniques of irradiation and dose determination, which are not, strictly speaking, radiation biology have been relegated to appendices.

Because this is an introductory text, general studies have been used as references. These direct the interested student to more complete treatment of the subject of a chapter.

New York, New York
Winston-Salem, North Carolina

DONALD J. PIZZARELLO
RICHARD L. WITCOFSKI

CONTENTS

CHAPTER · 1. General Introduction 1
2. Interactions with Matter 8
3. Direct and Indirect Action 20
4. Intracellular Responses: Targets 30
5. Intracellular Responses: Radiation Sensitivity ... 35
6. Intracellular Responses: Recovery 39
7. Intracellular Responses: The Influence of Oxygen, LET, and Dose Rate 44
8. Intracellular Responses: *In Vivo* Models 51
9. Intracellular Responses: Radiation Genetics 56
10. Effects on Tissues and Organs 67
11. Effects in the Total Organism: The Immediately Lethal Effects 71
12. The Hematopoietic Syndrome 76
13. The Gastrointestinal Syndrome 82
14. The Cerebrovascular Syndrome 89
15. Effects on Developing Embryos 93
16. Effects on Immunity 98
17. Treatment of Irradiated Organisms 103
18. The Late Somatic Effects 108
APPENDIX · A. Measurement of Radiation Dose 117
APPENDIX · B. Irradiation Techniques with External Sources of X or Gamma Rays 131
Index 137

Chapter 1

GENERAL INTRODUCTION

1.1 Introduction. Radiation biology concerns itself with the effects of ionizing radiations on or in living matter. Radiations other than ionizing radiations have biologic effects, but these comprise separate fields in themselves, and remarks in this book are confined to ionizing radiations only.

The effects ionizing radiations produce in living matter result from energy transferred from a radiation or radiations to one or more of the molecules of which cells are made. The primary effects of radiation, then, are *cellular,* and though there are often important ramifications at the tissue, organ, or organism level, these result from changes brought about in individual cells in the instant of radiation exposure. Only radiations that exchange their energy *in cells* have any promise of bringing about biologic effects; those interacting in extracellular substances are, as far as is known, inconsequential.

1.2 Interactions Between Radiations and Cells. The energy of radiation is transferred to cells when they interact with one or more electrons in the orbits of cellular atoms or molecules, or when one interacts with an atomic nucleus. In the

process, the radiation may lose all or part of its energy, and, of course, it is this lost energy that is absorbed in living matter.

1.3 Interactions Between Radiations and Orbital Electrons. What happens next depends on circumstances that will be explained more fully in Chapter 2. In brief, however, an interaction between a radiation and an orbital electron may result in either an ionization or an excitation. That is, the orbital electron may have been given enough energy in its interaction with the radiation to separate it from its atom or molecule (ionization) or only enough to move it to an orbit more distant from the nucleus (excitation). No matter, for both *ionized* and *excited* states are unstable, that is, they are high energy states with a relatively great potential for chemical reactions.

1.4 Cellular Chemical Reactions. Chemical reactions within cells are usually strictly and precisely regulated and may proceed only according to established patterns which, through the millennia of evolution, have been selected and offer good chances for cellular survival. Chemical reactions within cells are controlled,

1

therefore, by the overriding life process and should not occur at random (or at least outside prescribed pathways). One means by which this is prevented is that molecules lying adjacent to or near one another in cells do not react because many chemical reactions require some *energy* to initiate them. Once started, they may proceed without further added energy, but to begin, some exogenous source of energy is required. By selecting the molecules or molecular species that are given the energy to begin reacting, cells direct desirable reactions and prevent unwanted reactions.

Ionizing radiations interfere with this regulatory process. They supply energy needed to begin reactions by producing ions and excited states, and they do so practically at random. This is a key point. As a practical matter, all ionizing radiations are energetic enough to ionize or excite any molecule, provided they properly interact with one of its orbital electrons. On some occasions then, when cells are exposed to radiations, the radiations ionize or excite molecules and make possible chemical reactions between cellular molecules not intended to react or at sites not intended to be reactive.

1.5   Interactions Between Radiations and Atomic Nuclei.   Even if the interaction between radiation and matter is between a radiation and an atomic nucleus, similar results are possible. Sometimes the nuclei are set free from the molecular structure of which they are a part. This happens, for example, when high energy neutrons interact with hydrogen nuclei. These hydrogen nuclei, protons, then interact with orbital electrons of neighboring molecules to cause ionizations and excitations and raise the possibility of chemical interactions outside the metabolic framework.

Sometimes after interaction, atomic nuclei break apart. This happens when negative pi mesons interact with atomic nuclei. The resultant scattered nuclear particles and radiations then cause ionization and excitation in the ways already suggested.

One important result of radiation exposure, then, is the production of reactive states in cells. From this there follows a high probability of occurrence of chemical reactions which normally may be forbidden in metabolism.

1.6   Chemical Reactions and Living Matter.   But what of these chemical reactions? What part do they play in radiation effects on living matter? It will depend to a large degree on which of the many cellular molecules that happen to react. Not all cellular molecules are equally important to the life of cells and many, because they are present in excess over the minimum required, are expendable. This is not to say that some of the *species* of cellular molecules are unimportant; all are presumed indispensable. It is to say that not *every* molecule within each molecular species is vital, and in fact many are extra.

This expendability is not true for all species of molecules. There is a molecular hierarchy in cells, a kind of pyramid of molecular species in which the number of molecules in each species gets smaller as the pyramid is ascended. At the pyramid's apex is the compound, deoxyribonucleic acid (DNA). The molecules of DNA are not thought to be present in great excess, and in some cases there may be no excess at all. Beyond its relative scarcity, this compound is exceedingly important because it encodes the information cells require for metabolism, *directs* cellular metabolism and, therefore, all cellular functions. The *functions* cells carry out, both for themselves and for other cells, are the direct result of the kind of DNA they have. Furthermore, the functions cells have derived directly from the structure of their DNA. That means the proper *functioning* of most cells is the direct result of a particular, specific structure of their DNA.

1.7 Chemical Reaction Alters Cellular Function. Chemical reaction between any two molecules (or atoms) results in (1) the loss of at least part of the identity of each of the starting molecules and (2) the formation of a new compound with a new set of chemical properties: some may be like, but some will be unlike those of the parent molecules. In living matter this may lead to the following: (1) Because the starting materials are lost during reaction (they become the product), whatever *functions* these may have carried out for the cell may be lost or impaired. (2) Because the product of the chemical union has a new set of properties derived from its new structure, it is possible new functions or activities may appear.

1.8 Targets. Is every cell deficient in some previously performed function after irradiation? Does every cell have some new function after irradiation? No, and for the following reasons: (1) As shall be seen in Chapter 2, not every radiation interacts in every cell. Radiation–matter interactions are a matter of chance, and regardless of how large a dose of radiation is used, there is always a chance any given cell will not sustain an interaction at all. (2) Not all molecules are actually used in metabolism. As previously described, many are extra, part of the excess with which cells are equipped. When such molecules chemically react with each other, it does not matter (unless the product happens to be a poison) because they are not to be functional in the first place. (3) Even molecules that enter into metabolism and are thus functional are often transient figures. That is, they have relatively short lives and are replaced by new ones of the same kind. Loss of function of such molecules, unless it occurs in large numbers of them, may not be detectable, and, even when large numbers are involved, may be transient.

Radiation interactions that produce permanent changes or effects are probably uncommon events, and from the point of view of producing permanent, detectable biologic effects, most absorbed radiation energy is wasted. Some, however, is not. Cells, after all, *are* damaged by irradiation, and this damage can only come about by radiation-related changes being produced in some cellular structure or structures. Many believe such structures, which have been called "targets," to be molecules of DNA. Permanent structural changes in DNA are likely to alter its function, in many cases to create a situation in which cells may lose the ability to perform one or more functions permanently. The *loss of function* on the part of cells, after radiation-related events in some key or target structure, is what is meant by "radiation effect" or "damage." Such losses are most likely to occur if DNA is altered. Most functions of cells are important and have been selected during evolution as good and necessary for life. When one or more are lost, we can say a cell is damaged because its *viability* will be impaired. If it meets a situation in which the lost function is needed, and if that situation is life-threatening, its ability to survive, its viability, will be impaired. That is, in fact, one of the important biologic effects of radiation: life-shortening, a reduction in the *expected* or *average* life span in an irradiated population due to radiation-related changes in molecules of cells.

Many investigators think that DNA may not be the only important target and that other cellular molecules controlling particular functions may be targets for the specific functions they control. Actually a good case can be made for considering the whole cell as the target of radiation because all parts of cells are interdependent and the life of the whole depends on the proper functioning of all of them. Damage (from any source, not only radiation) to any cellular part endangers cellular life. However, repair or replacement of damaged parts of cells can often be carried out, usually out of the resources mobilized by the injured cell. Therefore, radiation damage done anywhere in the cell involves the whole cell; the whole, in this instance, cannot really be separated into its parts.

This intimate interrelationship is well demonstrated as the events following the transfer of energy to cells (ionization and excitation) unfold. If the target theory is accepted (that damage to or loss of particular cellular functions is the result of radiation-related changes brought about in targets, which are important molecules controlling those functions), then it follows that the changes in the targets may be brought about either *directly* (radiation transfers its energy directly to the key molecule, ionizing or exciting it) or *indirectly* (radiation transfers its energy elsewhere in the cell, but even so it results in a change in the target or key structure). The latter, known as indirect action, is an important process in cellular radiation effects. Since ionizing radiation is, by definition, capable of ionizing or exciting *any* molecule, it will, when passing through cells, interact most often with the molecules present in the largest numbers. In most cells, these will be molecules of water because there is more water than anything else in cells. But *individual* molecules of water have a very low *probability* of being targets. There is a vast excess of them, and they control or produce no specific function.

1.9 Direct and Indirect Action. By the same reasoning, the random nature of radiation interactions means there is a small probability of direct action (ionization or excitation) on target molecules. Whether target molecules are DNA or some other species of molecule, they will be organic molecules. Because cells consist of approximately 70 percent water, 7

out of 10 radiation interactions will occur in cellular water. The remaining 3 in every 10 interactions may occur in any of the molecules present, organic or otherwise. For this reason, interaction of radiation and organic molecules is uncommon, and the interaction with a specific organic molecule will, depending upon radiation dose, be quite rare. DNA or other targets, therefore, have a low probability of being ionized or excited when cells are irradiated, but this does not mean it can be changed only by this kind of direct action. Energy exchanged in water or any other molecule can affect other organic molecules and change them. Ionized water molecules can give rise to reactive, diffusible entities called "free radicals," which are capable of reacting with and becoming part of other molecules. When this happens, the *structure* of molecules with which the free radical has interacted will be changed. At the least the free radical will have joined, at some location, with and become part of the molecule. Since structure and function are inextricably linked, this change in structure (depending on how great a change it is) leads to the probability that the function the molecule performed for the cell may be lost.

Although this example of the indirect action of radiations has been confined to molecules of water, ionized and excited states may lead to free radical formation. The free radicals, in turn, may diffuse in the cell, and target molecules suffer a risk of change caused by radiations, even when they have not been directly involved in any radiation interaction.

1.10 Repair or Replacement. From the preceding general overview, then, it can be seen that the response of cells to irradiation will depend on whether cellular molecules are ultimately altered by the experience and, if so, which of them are changed. If important molecules are changed, a loss of function and perhaps a change in behavior on the part of irradiated cells is to be expected, and it is this loss and change that are known as "radiation effects." Specific radiation effects depend not on the kind of radiation used but on the cellular molecules changed and whether the change is permanent.

Since living matter has evolved within an environment in which radiation has always been present, it seems reasonable to expect that means to counteract or neutralize radiation effects would also have evolved. The internal environment of living cells is composed of a fine balance of functions, all carried out by its molecules. The changes wrought by radiations (or other agents) which cause *loss* of function have a high probability of being detrimental. Each cellular function is presumed necessary and may be needed at some time in the life of the cell or the organism of which the cell is a part. Loss of function, then, is loss of a component which endangers the viability of the whole.

Cells, therefore, have developed or evolved at least two means of coping with irradiation. One is the regaining of, or replacing, parts which have been damaged by radiation. The other is to reduce the probability of interactions between free radicals and molecules not touched by radiation. The result is two-fold. The chance that key molecules will be damaged is reduced because free radical reactions are lessened, and the detriment cells suffer when their key molecules are changed is lessened because they are repaired and restored to their pre-irradiation condition. The probability of loss of viability or even death following radiation events is thus reduced.

1.11 Effects in Tissues and Organs. Except for single cell systems such as bacteria, unicellular organisms, and cells in certain kinds of tissue culture, most cells are part of organisms. Within an organism each will be among many in a tissue which itself is part of an organ. When total organisms are irradiated, the effects that follow are the result of radiation energy exchanged into cells, the cells of its tissues and organs. When parts of organisms are irradiated (as occurs during exposure for diagnostic and therapeutic studies using radiation) the effects that follow for the entire organism and for the portion irradiated are again the result of events that have taken place in the cells of the tissues irradiated.

The actual result depends on a number of factors: the amount of radiation given, the kind of radiation used, the age and kind of tissue irradiated. The tissues and organs of children and fetuses are more sensitive to radiation than those of adults. Also, differentiating tissues and organs are more sensitive to radiation than those already differentiated.

But, within the same organism, the *response* of various tissues and organs to radiation also varies. The response of tissues and organs seems dependent on what has come to be called the *kinetics* of cells in those tissues or organs. Briefly, cell kinetics refers to the complex of tissue or organ activities, which includes generation of new cells in the reproductive compartment of the tissues to replenish the loss of aged or spent cells from the functional compartment. In certain tissues, functional cells have a relatively short life and must frequently be replaced. Because of that, the generative or "stem cell" compartment of such tissues is quite active, and there is much cell division in the compartment. Also, many cells will be differentiating as they specialize and progress to the final stage of their lives, the functional state. During differentiation, the cells of a number of tissues and organs reproduce, and the combined activity of stem and differentiating cells is a tissue or organ in which there is a great deal of cell division.

The response or sensitivity of cells

depends in part on whether they are in division. Dividing or reproducing cells are, on the whole, more sensitive to radiation than those in the interphase or functional phase of life. Therefore it follows that tissues having a rapid turnover of cells also will be sensitive to radiation. Many of their cells will be in division and will, in consequence, be radiosensitive so that the tissue as a whole is radiosensitive.

This sensitivity is in contrast to that in tissues in which the cell kinetics are such that there is little turnover. In these, the functional, mature cells have long lives, and there is little need of replacement from the stem or generative compartment. The result is a tissue or organ in which there is little reproductive activity, little differentiation, and relative insensitivity to radiation.

The response of tissues and organs to radiation, however, is not limited to their radiosensitivity. That factor, partly dependent on mitotic activity, determines the number of cells to be killed and the degree of damage done to individual cells by given radiation doses. The rate at which a tissue or organ recovers from radiation exposure is also a factor in its radiation response. The recovery of a tissue and organ depends upon the capacity of individual injured cells in the tissue or organ for repairing radiation damage and the rate at which damaged cells are purged from the population and replaced by undamaged ones. This replacement is known as "repopulation." This, too, depends in large measure on the kinetics of cell activity in tissues or organs. Much lethal damage done in cells by irradiation is not expressed except if such cells divide. Irreparable damage to DNA and/or the chromosomes may not kill cells until they divide. At division, however, such cells die, and radiation damage is then expressed. Tissues or organs with rapid cell renewal or turnover quickly purge themselves of damaged cells be-

cause of the high probability of division in given cells of the tissue. They are quickly replaced by the descendants of cells which either were not damaged by irradiation or which were able to repair their damage. Radiosensitive tissues, then, lose a lot of cells after irradiation and have many badly damaged ones. But soon, owing to repopulation, such tissues are purged of their damaged cells and refill their ranks with undamaged cells. Radioresistant tissues, on the other hand, probably carry damaged cells for long periods because they repopulate very slowly.

The contrast between radiosensitive and radioresistant tissues is this: Those rapidly renewing lose many cells at irradiation but quickly renew through repopulation; those renewing more slowly lose relatively few cells at irradiation but recover less quickly through repopulation. Sensitive cells are easily damaged and easily replaced; resistant cells are hard to damage and to replace. Radiosensitive tissues or organs usually respond to radiation quickly, whereas resistant ones show no immediate response. Resistant tissues, however, may demonstrate their injury later if damaged cells are suddenly eliminated owing to a demand for a large number of new cells in those tissues. Demand for large numbers of cells in tissues or organs with concomitant increase in cell division may occur if there is an injury in which substantial numbers of cells are lost and must be replaced.

The response of total organisms, in turn, depends on the extent of the body irradiated, the tissues and organs irradiated, and the degree of response of those tissues. If radiation is local and comprises a small circumscribed region, only the tissues irradiated are likely to be affected. However, if irradiation is more extensive, comprising, for example, one third to one half of the body or more, then a large number of tissue and organ systems will be affected, and the whole body responds. What happens is that the function of various exposed tissues or organs becomes impaired and they perform at a level that is less than normal. The fate of the organism depends on which of its tissues are affected, but if the deficit is in vital tissues or organs and the deficit in function is very great and persists for a long time, the organism may die. Total body radiation, therefore, can be fatal and is so when the function of one or more vital organs is suspended too long.

Chapter 2

INTERACTIONS WITH MATTER

2.1 Introduction. The effects ionizing radiations produce in living matter are the result of energy transferred from these radiations to one or more of the atoms or molecules of which cells are made. The mode of energy loss which is characteristic of ionizing radiations involves either a direct or an indirect transfer of energy to atomic electrons resulting in either excited or ionized atoms and molecules. As pointed out in Chapter 1, these ionized or excited states are the first events in the chain leading to biologic effects.

2.2 Atomic Structure: The Nucleus and the Electron Cloud. An understanding of the structure of atoms is important, since the interactions of radiation and matter occur at the atomic level. An atom has a small, dense, positively charged nucleus which includes practically its total mass. The nucleus contains uncharged neutrons and positively charged protons which are of almost equal mass (approximately 1 atomic mass unit) tightly packed together. This is surrounded by a diffuse cloud of small, rapidly moving electrons which rotate about it at relatively great distances from it. Electrons have little mass (it would take almost 2000 electrons to equal the mass of a proton or neutron), and because of its diffuse nature, the cloud occupies a great deal of space and gives the atom most of its size. Although the electron cloud does occupy a great deal of space, the "diffuseness" means that there is much space between electrons and between electrons and the nucleus so that atoms are mostly "empty space." For this reason, ionizing radiations may pass through many, many atoms before they chance to interact with an electron. Interactions between ionizing radiations and atomic electrons occur on a random basis, whenever a radiation comes close enough to or collides with one of the widely spaced orbital electrons.

2.3 Electrons. Each electron carries a unit negative charge, equal and opposite in sign to the charge on the nuclear proton. The electrons rotate about the nucleus at specific distances in orbits and are held in these orbits by the attractive force between the negative electrons and the positive nucleus. Therefore, electrons in orbits close to the nucleus and to its concentration of positive charge are bound very tightly in the atom. Those in orbits more distant from the nucleus, and therefore more distant from the attraction of its positive charge, are bound less tightly in the atom. There is also a relationship between the number of protons and electrons. In the *stable* atomic configuration the *atom is electrically neutral,* since the number of positive protons equals the number of negative electrons.

2.4 Ionization. The principal means by which ionizing radiations dissipate or transfer their energy to matter is by the ejection of orbital electrons from atoms. This process, in which one or more electrons are removed from an atom, is called *ionization,* the ionized atom and the ejected electron constituting an *ion pair.* The energy dissipated per ionization (the *average* energy required to remove an electron from an atom) is about 34 eV.* Because the energies of ionizing radiations are usually much greater (a factor of 1000, or more) than the energy required to remove an atomic electron, electrons which are removed from atoms during ionization may be ejected with high speeds (high energy). This energy is lost in turn through high-speed electron-orbital electron collisions causing a number of other ionizations along the path taken by the electron ejected at the original ionization. Consequently, ions produced during irradiation are not distributed at random but are concentrated along the tracks of ejected electrons. Ultimately, the ejected electrons come to rest, their energy expended along well-defined tracks within the medium.

2.5 Excitation. Not every interaction of an ionizing radiation in matter need result in ionization. Excitation, the most important mode of energy dissipation by ultraviolet light, accounts for a significant

*The electron volt (eV) is a very small unit of energy equal to the energy acquired by an electron falling through a potential difference of one volt.
10^6 eV = 10^3 KeV = 1 MeV

percentage of the energy dissipated by ionizing radiations in tissue (approximately 20 percent of the energy of a high speed electron). Excitation involves electrons in the outermost orbit or shell called valence electrons because they are responsible for chemical combinations with other atoms. Since they are farthest from the nucleus, they are quite loosely bound. If a small amount of energy is given to them, they become excited; that is, they move even farther from the nucleus to higher energy levels. Excitation can bring about changes in the chemical forces which bind atoms within molecules (the C≡C bond is relatively strong, yet it is only 3.9 eV). This change of forces may or may not lead to a regrouping of affected atoms that will result in a different molecular arrangement. Thus, the excitation of external electrons is an indirect method of inducing chemical changes by supplying needed energy and by breaking bonds between atoms and molecules.

2.6 Types of Ionizing Radiations. The characteristics and origins of the more important types of ionizing radiations are given in Table 2.1. Ionizing radiations fall into two general categories: those which have mass (alpha, beta, protons, etc.) and those which are only energy (x rays and gamma rays). Those with mass may be charged (alphas or protons) or uncharged (neutrons), but those without mass are never charged.

Alpha Particles. The alpha particle is a large highly charged particle usually associated with the decay of the heavy naturally occurring radioactive elements (such as radium) and is made up of two protons and two neutrons. Because of their large size and charge, alpha particles are relatively slow moving and lose their energy along short, straight tracks of exceedingly high ion density. Even very high energy alpha particles are able to penetrate only up to a few microns in tissue.

Table 2–1. Types of Ionizing Radiation

Type	Mass	Charge	Description	Produced by
Alpha	4	+2	Doubly ionized helium atom	Radioactive decay primarily of heavy atoms
Beta (negatron)	0.00055	−1	Negative electron	Radioactive decay and betatrons
Beta (positron)	0.00055	+1	Positive electron	Radioactive decay and pair production
Protons	1	+1	Hydrogen nuclei	Van de Graaff generators and cyclotrons
Negative π mesons	0.15	−1	Negatively charged particle with a mass 273 times an electron	Accelerators
Heavy nuclei	Have a range of masses	Have a range of charges	Any atom stripped of one or more electrons and accelerated will be an ionizing particle. Deuterons and carbon atoms are examples.	Accelerators
Neutrons	1	0	Neutron	Atomic reactors, cyclotrons
Gamma rays	0	0	Electgromagnetic radiation	Radioactive decay
X rays	0	0	Electromagnetic radiation	X-ray machines and from the rearrangement of orbital electrons

Beta Particles. Beta particle emission during radioactive decay is of two kinds: emission of positive electrons (positrons) or of negatively charged beta particles (negatrons). In addition to radioactive decay, high speed negative electrons can also be produced by betatrons. Ionizations produced by beta particles as they traverse matter are much farther apart than those of alpha particles, so that beta particles travel farther through matter before coming to rest. A high energy electron (about 1 MeV) will have a range in tissue of about 1 cm.

Gamma and X Rays. These electromagnetic radiations are produced by radioactive decay or by x-ray machines.

They have neither mass nor charge; they are just energy. These "bundles of energy," or *photons*, are capable of penetrating through considerable thickness of most materials, including tissue, because their interactions with atomic electrons are so sparse. Their ability to penetrate matter is usually given as their half-value layer in aluminum, copper, or lead (the thickness of the half-value layer is a material which will reduce the intensity of a beam of photons to one-half its original intensity).

Protons and Deuterons. Protons (hydrogen nuclei) and deuterons (heavy hydrogen nuclei) are, in the main, produced in cyclotrons, proton synchrotrons, and van de Graaff generators. Their ionizing tracks are straight because they are undeflected by the less massive electrons with which they interact. Nevertheless, their direction can be radically altered by occasional interactions with atomic nuclei. They are intermediate between alpha particles and beta particles in their ability to penetrate tissue.

Neutrons. Fast neutrons (uncharged particles with a mass of 1) are obtained from a variety of sources such as cyclotrons, reactors, and by the bombardment of beryllium with alpha particles. Fast neutrons do not ionize directly but knock out protons from the nuclei of atoms in the absorbing medium. Slow neutrons do not eject protons but are captured by atomic nuclei which in turn may be made radioactive. Because the probability of atomic interaction for uncharged particles is so small, neutrons are capable of penetrating considerable distance in tissue.

Mesons. Negative pi mesons are produced by high energy accelerators. These negatively charged particles are intermediate in mass between electrons and protons (273 times the mass of an electron or 15 percent of a proton mass). They lose energy through multiple atomic interactions, but because of their high energies, they are capable of penetrating a considerable distance in tissue. Near the end of their tracks, slowed mesons are captured in the orbits of an atom to form a mesonic atom which immediately disintegrates, liberating highly ionizing particles.

Heavy Nuclei. Any atom stripped of its electrons and accelerated will be an ionizing particle. Accelerators now available can accelerate nuclei of atoms such as carbon and nitrogen to very high energies. The large mass and charge confer upon these nuclei the capability of colliding with atomic nuclei and accelerating those nuclei (a reaction analogous to the production of high speed electrons).

2.7 Interactions of Radiation in Matter. *The Interaction of X and Gamma Rays.* X and gamma photons have no mass or charge. In their interactions with matter the energy of these photons is transferred by collision, usually with an orbital electron in an atom of the *absorbing* medium, the medium through which they are traveling. Following such interactions an electron may either have been moved to an orbit more distant from the nucleus (the atom is excited) or, more commonly, have been *ejected* from the atom (ionization) with high energy and at a high speed. The energy given *to* this electron will be dissipated *from* it as it moves through the medium; it will ionize and excite atoms with which it interacts.

Since the interaction of x and gamma ray photons with matter depends almost entirely upon direct collision with orbital electrons (and this is a relatively rare event) they may penetrate deeply into matter passing though vast distances in it without having interacted. Atoms are composed chiefly of unoccupied space, and the photon, having no charge and consequently no affinity for any of the atomic parts, may pass undeflected through the relatively vast intra- and interatomic spaces. In fact, there is some probability of penetration, even through the greatest thickness of the most dense matter, without a single interaction ever having taken place.

The probability that a photon *will* interact when it passes through any given thickness of matter will depend on a complex relationship between (1) the *energy* of the photon, (2) the *density* of the matter, and (3) in some interactions, on the *physical makeup* of the atoms of which the matter is composed. The two major modes by which the photons give up their energy to a medium through the ejection of orbital electrons depend upon photon energy. They are (1) photoelectric absorption and (2) Compton scattering.

Photoelectric Absorption: Low-Energy Photons. When a low-energy photon collides with an orbital electron the most likely result will be a transfer of all of the photon's energy to the electron. The photon disappears entirely and is thought of as having been "absorbed" by the electron.

The electron itself will be ejected from its orbit and from the atom. The process is, of course, ionization, and the result will be the formation of an ion pair. These ejected electrons, now moving at high speeds, move through tissue, producing ionizations among the atoms which they traverse and with which they interact. The probability of occurrence of photoelectric interactions decreases as photon energy increases. In soft tissue, photoelectric absorption is the predominant energy absorption process for photons having energies up to 100 KeV (those produced by x-ray machines operating at voltages up to 150 kilovolts).

The probability of photoelectric absorption depends not only on the energy of the photon but also on the *atomic number* of the atoms with which it interacts. There is preferential absorption of photons in material composed of elements having high atomic numbers (bone) when compared to material composed of elements having low atomic numbers (soft tissue). The consequence of this is that for the *same* exposure to lower-energy photons, bone will absorb more x or gamma radiation than will soft tissue.

Compton Scattering: Median Energy Photons. In contrast to photoelectric absorption, Compton scattering is a process in which higher energy photons interact with matter. In it the photons have only a *portion* of their energy absorbed in interacting with orbital electrons, but they are not *totally* absorbed and do not disappear. The process is confined, for the most part, to interactions with outer loosely bound electrons. Because these electrons *are* loosely bound, nearly *all* the

energy exchanged to the electrons will be in the form of electron kinetic energy. The photon itself will be deflected in the interaction; its direction will be changed. Because of this, it is said to have been "scattered." The products of the interaction are (1) a scattered, less energetic, degraded photon (some of its energy has gone to the electron it ejected), (2) a high-speed electron, and (3) an ionized atom.

The ejected electron, as in photoelectric absorption, will travel some discrete distance in matter, producing ionizations along its track. But in Compton interactions, there is also the degraded photon which may undergo two, three, or more additional Compton collisions before all its energy is finally lost in photoelectric absorption.

Since Compton interaction principally involves the most loosely bound electrons, it will also differ from photoelectric absorption because it will *not* depend on atomic number (the attractive force of the nucleus). And *because* only the most loosely bound electrons are involved, all of them will absorb about the same amount of energy from the photon; in effect, none goes to the atom. The Compton process is a random process, and the probability of interaction will depend only on the density of electrons (the number of electrons per gram of material). Most materials have nearly the same number of electrons per gram, so that it follows that Compton absorption per gram of matter is almost the same for all materials. This, too, is in contrast to the photoelectric effect where high atomic number materials, such as are present in bone, absorb more energy from the same dose of incident radiation than do materials such as soft tissue. For photons in the range of 100 KeV to 10 MeV, the Compton interaction is the most important photon interaction process in soft tissue.

Pair Production: High Energy Photons. The energy of the photon may be exchanged into matter by yet another mechanism which differs from those already described in two fundamental details: (1) it occurs exclusively with *high-energy* photons, and (2) the interaction is with the *atomic nucleus* and does not involve the ejection of orbital electrons.

Photons with energies greater than 1.02 MeV may interact with the electric force field of the highly charged nucleus so that their energy is converted to mass. The photon is changed into two particles, a *positive* and a *negative* electron. (The process can occur in reverse; mass can also be converted to energy. Energy (E) and mass (m) are related to each other according to the Einstein formula, $E = mc^2$, where c is the velocity of light.) Pair production cannot occur with photons whose energy is less than the rest mass of *two* electrons, a threshold energy of 1.02 MeV. Photon energy in *excess* of the threshold value will be shared as *kinetic energy* between the two newly formed electrons. If there is no excess of energy present, the electrons will immediately recombine ("annihilation") and are converted back to energy (2 photons of 0.51 MeV each).

In the instance where threshold energy is exceeded, the electrons will move away from the point of formation. Both will move through matter undergoing interactions with and ionizing other atoms in the substance until the excess kinetic energy is exhausted. In this way, energy is ultimately transferred from the photon to matter.

When the positive electron comes to rest, it will interact with an available negative electron. Annihilation will occur, the electrons will disappear, and the electron masses will be converted to two photons which share the energy.

Pair production is dependent upon atomic number. The initial photon interaction is with the force field of the nucleus. Larger nuclei, having more protons, will have stronger force fields, thus increasing the probability of the occurrence of this process.

The Relative Importance of Different Types of Photon Absorption. Predominance of one or more of these processes will depend upon the energy of the photons and the nature of the absorber. Table 2–2 summarizes the relative *preferential* energy absorption of bone and soft tissue when each of the photon absorption processes is predominant. For example, x-ray tubes operating at voltages up to 150 KV_p will produce beams resulting in high bone absorption; most of the interactions are of the photoelectric type. For x-ray machines operating between 150 and 250 KV_p, the predominance of the photoelectric effect diminishes, but bone will still absorb some extra energy. Photons with energies from about 1 to 10 MeV (cobalt-

Table 2–2. Relationship of the Primary Photon Energy Absorption Process of the Energy Absorption in Bone and Soft Tissue

Photon Energy	Primary Absorption Process	Preferential Absorption
Low	Photoelectric	Bone will absorb 5 to 6 times as much energy per gram as will soft tissue
Medium	Compton	Bone and soft tissue absorb essentially the same energy per gram
High	Pair production	Bone absorbs approximately 2 times as much energy per gram as soft tissue

60, gamma rays, and high-energy x-ray machines) lose energy exclusively by Compton interaction so that there is no preferential bone absorption. With more complex equipment such as betatrons which operate with energies from 10 to 30 MeV, pair production is predominant and preferential absorption by bone again occurs.

Interaction of Neutrons. Neutrons, since they are uncharged particles, are not affected by and cannot themselves affect charged objects (atomic nuclei or electrons) except by direct collision. Since their interactions must depend upon chance collisions with atoms, neutrons are able to penetrate great distances in matter of all kinds. Neutrons are classified by their only distinctive property—their energy. High energy neutrons are usually considered *fast;* low energy neutrons, *slow.* Fast neutrons lose energy mainly by collision with atomic nuclei. Because hydrogen atoms are the most numerous in tissue of average water content (tissues are 70 to 80 percent water), and since the average energy transferred from a fast neutron to a hydrogen nucleus is much greater than the energy transferred to any other nucleus, for practical purposes the major mode of energy loss of fast neutrons in soft tissue may be considered to be by the *ejection* of high-speed protons (hydrogen nuclei). The protons will have a variety of energies, depending upon the neutron energies, but all of them will be highly ionizing particles, which will lose energy rapidly along their tracks by ejecting atomic electrons.

Slow neutrons interact in matter mainly by the process of "capture." The uncharged, slowly moving neutron actually enters the nucleus of an atom in matter and loses its identity; it becomes just another neutron. The added neutron— and its energy— may put the atom in an unstable state; the atom may become radioactive and itself emit charged particles or gamma rays.

Interaction of Charged Particulate Radiations with Matter. The interactions of *charged* particles with matter are distinct because they have both mass and charge. These properties make possible interactions with matter not only by *direct* collision with electrons in the orbits of the atoms in the matter that they traverse but also by *interactions* (this can be attraction or repulsion) *between their charge and that of the orbital electrons.* But, the final result of the interaction of both electromagnetic and particulate radiations will be the same: that is, ionization, the production in matter of high-speed electrons.

The Influence of Particle Charge. The effect which one charged body has upon another is related to the *distance between them* and the *amount of charge* on each. As the distance between charged bodies increases and/or the quantity of their charge decreases, the effect they have on each other diminishes. The rate at which energy is lost along the track of the charged particle depends upon the *square* of the charge of the particle. Thus, a proton and an electron moving through matter at the same *velocity* would have the same rate of energy loss, but an alpha particle (with twice the charge of either the proton or the electron) moving through matter with the same velocity as either a proton or an electron would have a rate of energy loss four times as great. The more highly charged the particle, the more intense *its* electric force field, and the greater the likelihood of producing ionization of atoms along its track. This is because charged particles, in addition to colliding with orbital electrons, may also attract or repel them from orbit. The influence of particle charge is to increase the number of interparticulate interactions and hence produce more ion pairs per unit path length.

The Influence of Particle Velocity. Two particles with the *same* energy do not necessarily have the same velocity.

The kinetic energy (energy of motion) of any object is equal to ½ mv², where m is the rest mass and v is the velocity. For example, protons, because their rest mass is about 2000 times that of electrons, will be moving much slower than electrons when the kinetic energies of both are the same. The influence of particle velocity on interactions between charged particles lies in the fact that velocity will control the *length of time* the electric force field of the particle is exerted in any given place. The effect (attraction or repulsion of orbital electrons) by the force field of the charged particles will depend on how long the force is applied. The slower moving charged particles will produce more ionizations per unit path length (a higher "specific ionization") than faster moving ones. Two particles with the *same* kinetic energy may produce vastly different specific ionizations if their masses differ. For example, even though their energies are the same, a 1 MeV proton would have a velocity of only 5 percent that of light whereas a 1 MeV electron would have a velocity of over 90 percent that of light.

During each interaction, energy is lost by the particle and transferred to the medium. A loss of kinetic energy will inevitably result in a change in velocity; after each loss the particle will move more slowly. Since the probability of ionization is inversely related to velocity, such a change will mean an increase in ionization density and an increase in the rate of loss of kinetic energy. The ionization density of a charged particle reaches the highest point just before the particle comes to rest.

2.8   Linear Energy Transfer (LET). *Specific ionization* is defined as the number of ions formed per unit length of path length of particle and takes into consideration only the energy transferred to the medium by *ionization*. But energy will also be transferred to the medium by

excitation. At present the full biologic significance of excitation is not known, but a substantial amount of energy is transferred to tissue this way. The process is likely, therefore, to be very important. A unit that has been devised to account for *all* the energy liberated along the path of an ionizing particle, irrespective of the mechanism, is linear energy transfer (LET). LET is the energy released (usually KeV) per micron of medium (tissue) along the track of any ionizing *particle*.

Since *rate of loss of energy* by ionizing particles will be affected by the velocity and the charge on such a particle, a relatively slow-moving highly charged particle will have a high LET. A faster-moving particle and/or one with a lesser charge will have a much smaller LET. Values of LET are given in Table 2–3 for a number of different radiations. Biologic damage is related to LET. In a general way, particles with high LET are *more likely* to produce change in a given volume of living matter because their ionizations are produced close together in comparison to low LET radiations, whose ionizations are produced relatively far apart.

The increase in degree of biologic effect with increasing LET does not, however, continue indefinitely. With radiations of extremely high LET, *more* energy is transferred to the system than is needed to produce even the maximum biologic effect—death. Such radiations will "overkill," that is, they will deposit more energy in cells or tissues than is required to inactivate them.

Particle Tracks in Matter. LET is not a static or constant value but will be different even for the same particle over different portions in the track. This is so because, though charge on a particle is a constant factor, the velocity will be continually changing (decreasing) all along the particle track. Each interaction (excitation or ionization) involves a *loss* of energy from the particle and a concomi-

Table 2–3. LET Values of Ionizing Particles (1, 2)

Particle	Charge	Energy (MeV)	LET (keV/μ)
Electron	−1	0.001	12.3
		0.01	2.3
		0.1	0.42
		1	0.25
		200 KVp x rays*	0.4–36
		Cobalt 60 rays*	0.2–2
Proton	+1	Small	92
		2	16
		5	8
		10	4
Alpha	+2	Small	260
		3.4	140
		5	95
Neutron†	0	2.5	15–80 (peak at 20)
		14.1	3–30 (peak at 7)

*This applies to the LET of secondary electrons ejected by photons.
†Neutrons produce no ionization directly in passing through tissue. These values are
for the protons ejected in collisions.

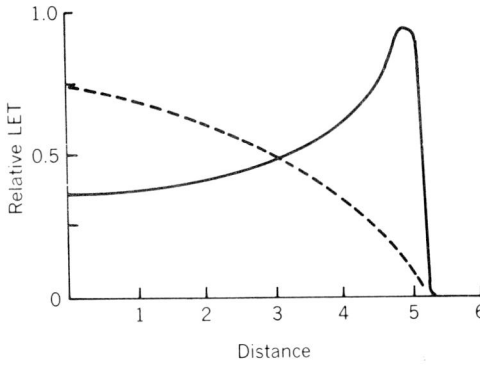

Figure 2.1. The rate of dissipation of energy of a charged particle as it moves along its path. The rate increases gradually as the particle slows down but dramatically increases (the Bragg peak) just as the particle is coming to rest. The broken line indicates the residual energy of the particle as it moves along the track.

tant deceleration. As a result, LET gradually increases along a particle track with a dramatic increase occurring just before the particle comes to rest. This peak in the rate of energy dissipation is called the "Bragg peak" (Figure 2.1). When a particle does comes to rest, it usually acquires electrical neutrality. Positively charged particles will acquire electrons (the pro-

ton will become a hydrogen atom; the alpha particle will become a helium atom), and high-speed electrons, when they come to rest, drop into atomic orbits.

Particles with different amounts of charge will produce different tracks. Highly charged particles interact frequently; consequently, they have a high LET, and the ionizations along their tracks are very dense. Particles with lesser degrees of charge are sparsely ionizing; they have a lower LET.

The interaction of any particle in matter is a random affair. This is true because the atoms of all matter are in constant motion and, except for the rigid structure of crystals, are randomly spaced. In addition, the electrons of each atom are also in constant motion. Thus, an ionizing particle traveling through matter will encounter orbital electrons on a chance basis. For particles having a very high LET, the *randomness* of interaction will have little biologic importance. The interactions of these particles are so closely spaced that, even when the particles pass through a space as small as the tiniest cellular compartment, it is highly likely that some energy will have been left there. In contrast to this, sparsely ionizing (low LET) particles interact infrequently enough so that it is even possible for them to pass through cells or their important substructures without having interacted at all.

Whether the *track* of a particle is altered by its interaction with an orbital electron will depend upon the particle's mass. Large massive particles (such as alpha particles) have straight tracks; they are not deflected in their interactions with the minute electrons. Their direction is changed only when they collide (it will be a chance event) with an atomic nucleus. Electrons, on the other hand, have the direction of their track changed at nearly every interaction. Their tracks, usually described as tortuous, are the result of deflections which come about from the interaction of two bodies of equal mass.

Not every electron ejected by an ionizing particle will have the same kinetic energy; particles do not lose the same amount of energy at each interaction. Some electrons are scarcely removed from their orbits; others are given large amounts of kinetic energy and move for appreciable distances through matter. Some of these energetic "secondary" electrons may have sufficient energy to produce a track of their own which will appear as a branch from the track of the primary ionizing particle (Figure 2.2).

Such spurs or branches are known as delta (δ) rays. They themselves will have a LET that will be different from the primary particle as well as from other delta rays formed from the same ionizing particle. Their major importance in radiobiology is in the fact that they distribute the energy of the primary particle to regions *outside* the primary particle track.

The distribution of ionization from electrons usually occurs in clusters that have a fairly random distribution. The *number* of ions in a cluster will vary, but on the average there will be approximately 3 in each. Assuming the value of 34 eV/ion pair to be approximately correct for soft tissue, an average of 3 ionizations per cluster would mean that approximately 100 eV is deposited at each point of interaction along the track. The effect of decreasing velocity (with increasing LET) is to cause increasing numbers of ions per cluster. Finally, a rough estimate of the distribution of energy from a fast electron (1 MeV) in water has been given by Johns:[1] 20 percent excitation, 40 percent ion clusters, and 40 percent delta rays.

2.9 Radiation Dose. Nearly all radiation effects are dependent upon the amount of energy absorbed (the dose) so that it is imperative to be able to measure dose in units that are not ambiguous. Units of radiation quantity should be specifically defined and universally accepted so that physical irradiation techniques and the results obtained can be reproduced exactly.

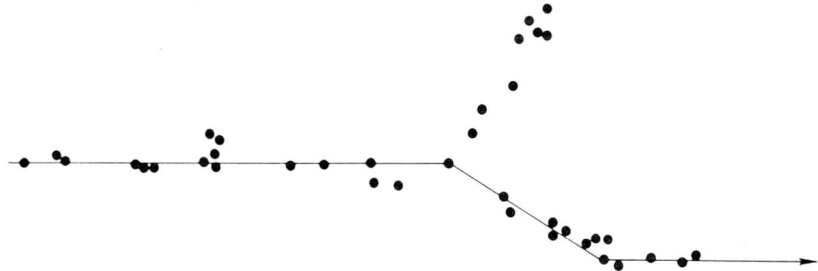

Figure 2.2. Representation of the distribution of ionization along the track of a charged particle. A delta track or "spur" is shown branching from the primary track.

Two units have been adopted to express the measurement of radiation dose: the roentgen and the rad. The roentgen (R) is a unit related to the ability of x or gamma rays to ionize air, and the rad is a measure of energy (100 ergs) absorbed per gram for any type of radiation. The roentgen is restricted solely to *exposure* and the rad to *absorbed dose*. In keeping with this concept, when the term *dose* is used in this text, it should be understood to mean *absorbed energy*. The terms are defined and details and techniques of exposure and dose measurements are found in Appendix A.

2.10   Relative Biologic Effectiveness (RBE).   The amount of biologic damage produced by different ionizing radiations may vary by a factor of 10 or more for the same rad dose (the same quantity of *absorbed* energy). The reason for this disparity is that the rad is a macroscopic quantity, which, under ordinary conditions, represents the amount of energy absorbed by a mass of tissue much larger than a living cell. On the microscopic level, the amount of energy absorbed by individual cells in a volume of tissue given a dose of 1 rad may vary tremendously, so that some cells will be completely spared while others may be lethally damaged. For example, if the dose were delivered by x rays, the relatively sparse ionization density along the tracks of the secondary electrons would result in a fairly uniform deposition of energy. Thus, although many cells might absorb energy, it probably would be insufficient in many instances to irreversibly damage cells. On the other hand, if the *same* dose were delivered with alpha particles, the energy distribution would be "spotty," and the dense ion tracks might be lethal to every cell that was hit. The same amount of energy would be absorbed (the same rad dose), yet more cells would die in the tissue irradiated with alpha particles than in the tissue irradiated with x rays.

Measures or estimates which compare the effectiveness of two types of ionizing radiation to produce a particular biologic effect are called the relative biologic effectiveness or RBE. In most RBE determinations, the standard radiation against which others are compared is medium energy x rays (200 KV$_p$, with a LET = 3 KeV/μ), at a dose rate of about 10 rads/minute. RBE is usually expressed as a ratio: RBE = D_{std}/Dx, where D_{std} is the absorbed dose from the standard low LET radiation and Dx is the dose (required for the same effect) from the comparison radiation. Usually, the distribution of deposited energy along individual tracks will determine the effectiveness of a radiation for producing a given result. For most radiations, the RBE increases as the LET increases (Figure 2.3). RBE appears to rise to a maximum at 100 to 200 KeV/μ and then falls again. This decrease in RBE with extremely high LET results from a transfer of more energy to the system than is needed to produce the maximum effect—usually cell death.

RBE often depends also upon the dose used, as well as the LET of the radiation. Specifically, the RBE is usually higher

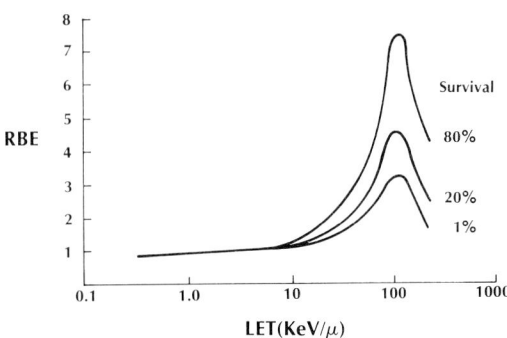

Figure 2.3.   An example of variation of RBE with LET for a particular experiment. The end point for biologic effect was the ability of cells growing in tissue culture to survive irradiation of differing LET values. The percentages in the figure are for 80 percent, 20 percent, and 1 percent survival. Note the dependence of RBE on the dose (higher RBE for smaller doses). The fall in RBE at higher LET values is a result of the transfer of more energy than necessary to inactivate cells. (Redrawn from Barendsen and Walter.[2])

for *smaller* total doses. This is clearly seen in Figure 2.3, where the lowest dose (which gives 80 percent survival) has a higher RBE than the higher doses which produce lower survival (20 percent and 1 percent survival).

RBE has the most meaning when it is applied to small volumes of test materials, such as bacteria or cells in tissue cultures. The application of the RBE concept to large animals has given rise to considerable confusion. There is no single RBE for all biologic end points for any given radiation. For instance, in the mouse, RBE values vary from tissue to tissue (the RBE for cataract formation may be higher than that for testicular atrophy). In addition to this, RBE also may vary with dose rate, the manner in which the dose is measured, and the distribution of the dose in the tissues.

When x rays and gamma rays constituted the only available types of ionizing radiations, the *quality* was used to describe the energy and penetrating power of the radiation. Quality was usually expressed in terms of the half-value layer in copper or aluminum. With the advent of other types of radiation it was realized that the biologic effect per unit dose absorbed depended upon the type of radiation used. The term *quality* or *quality factor* became a description of the radiation as it affects the biological response; two radiations have the same quality if the biological response per rad is the same.

SUMMARY

1. The atom consists of a massive portion, the nucleus, which contains the positive protons and uncharged neutrons.
2. The nucleus is surrounded by negatively charged electrons which are found in discrete orbits at distances from the nucleus.
3. In the stable atom the number of protons and electrons is equal—the atom is electrically neutral.

4. Radiations may have sufficient energy to remove an electron completely from an atom and produce an electrical charge (ionization) or perhaps only to move an electron to an orbit farther from the nucleus (excitation).
5. These processes, particularly ionization, are responsible for biologic damage produced by ionizing radiations.
6. X rays and gamma rays lose their energy principally by ejecting high-speed electrons from atoms. These secondary electrons dissipate their energy by the production of further ionizations and excitations.
7. LET is a measure of energy lost per unit of path length (KeV/μ) for particles as they lose energy in matter (tissue).
8. LET varies as the square of the charge of a particle and inversely as its velocity.
9. Charged particles such as π mesons, protons, alpha particles, and heavy nuclei lose the majority of their energy through electron interaction but may also interact with nuclei.
10. High-speed neutrons lose their energy in tissue primarily by the ejection of high-LET protons from water.
11. The roentgen is the unit of exposure for x rays and gamma rays.
12. The rad is the unit of dose which relates to energy absorbed.
13. RBE relates biologic effect to energy absorbed (rads to biologic effect).

Text References
1. Johns, H. E., and Cunningham, J. R.: The Physics of Radiology. Springfield, Ill., Charles C Thomas, 1969, pp. 678–679.
2. Barendsen, G. W., and Walter, H. M. D.: Effects of different ionizing radiations on human cells in tissue culture. Radiat. Res., *18*:106, 1963.
3. Radiation quantities and units. ICRU, Report 10a, NBS Handbook 84, 1962.

Chapter 3

DIRECT AND INDIRECT ACTION

3.1 Introduction. Changes in or losses of cellular functions most likely come about after changes produced as a result of irradiation in certain of the molecules of which cells are composed. Certainly a number of cellular functions are known to be localized in or on given subcellular structures (the genes are on or are part of the chromosomes; certain enzymatic functions are confined to mitochondria, to cite two examples). It seems reasonable to suppose that change in, or loss of, many cellular functions probably comes about after fundamental radiation-produced changes in the structure and function of the molecules which carry them out or control them. The cellular site or organelle or unit that controls or carries out a cellular function may then be thought of as a "target" for radiation. Unless radiation brings about a change in certain sites or targets, given cellular functions will probably persist after irradiation. The molecules of which such sites or targets are composed are likely to be relatively complex, organic ones and not the simpler inorganic compounds that are found in many cells. A change can be brought about in any given molecule by radiation if the molecule is directly ionized or excited and subsequently reacts with a nearby molecule or is otherwise altered functionally or structurally because of the experience. This action of radiation is called "direct action." But there is an alternative. Molecules may be changed if they react with other molecules that have been directly ionized or excited by radiation or the products of some other molecule that was directly ionized or excited. This phenomenon has been called the "indirect action" of radiation.

In cells, indirect action is likely to be the more important action of radiation because those important target molecules controlling given functions are probably few in number compared to the total number of molecules present, and direct action, a random event, is likely to be a rare event. Indirect action is probably more common.

3.2 Energy Transfer in Complex Systems. If a complex system (one consisting of more than one kind of molecule) is irradiated with any of the ionizing radiations, then, because the nature of the energy exchange is random, ionization is most likely to occur in those kinds of molecules that are present in the largest

20

number. This is not to say that the other kinds of molecules in the mixture will be unaffected by the radiation. They will be ionized in the proportion in which they are present. The total amount of radiation given a mixture will, of course, determine the precise status of the mixture in the instant after irradiation, but, unless a dose so large as to ionize every molecule in the mixture is given, the following situation will result. Some molecules of every type will have been ionized (the greatest number of *ionized* molecules will be found among those that are present in the greatest number), and some of every type will be left intact. The precise ratio of ionized to intact molecules is determined by the total dose of radiation. But, unless the dose was large enough to ionize all molecules, some molecules of every type will escape this kind of change.

However, this does not mean that *all* the intact molecules will escape *radiation-related* change. Changes (although not necessarily ionizations) may be brought about in them by the *irradiation products* of a molecule which has, itself, been ionized. The energy of an ionizing particle can be transferred to one of the intact, un-ionized molecules from a molecule that has been "hit" (ionized).

There are, then, two quite different mechanisms by which chemical changes in molecules may be brought about by ionizing radiation. One is *direct action*, i.e., a molecule is ionized or excited (see Chapter 1) by the passage through it of either an electron (secondary to electromagnetic radiation) or another ionizing particle. The other is *indirect action*. The changed molecule has not itself been ionized or excited by a particle; no ionizing radiation passes through it. It is changed, however, because it received the energy of an ionizing particle by transfer from another molecule that *has* been ionized through the direct action of radiation.

3.3 Observations in Solutions. The difference between the two modes of action—direct and indirect—is especially striking when the irradiation of solutions is studied. A solution, which is a type of mixture, will contain molecules of solvent and solute (more than one kind of solute can, of course, be present in a single solution). Inevitably, the molecules of solvent will far outnumber those of the solute, and, if the solution is irradiated, because energy exchanged from ionizing radiation is random, most of the energy will be transferred in the solvent. Although more molecules of solvent than of solute are present, solute molecules will also be ionized, but, compared to the energy exchange in the solvent, in lesser proportion. The majority of changes made in the molecules of solute will be due to energy transferred to these molecules by the irradiation products of directly ionized solvent molecules.

3.4 The Analogy of the Cell to a Solution. It should be kept in mind that cells (and, therefore, living things which are composed of cells) are extremely complex mixtures or solutions. Water is the solvent, and the chemical reactions that make up the process called metabolism take place in it. There is, of course, some variation, depending upon the tissue, but on the average, cells consist of about 70 to 80 percent water. The cell molecules (proteins, carbohydrates, nucleic acids, inorganic substances) are myriad, but they are either dissolved in or suspended in a watery medium. When cells or tissues are irradiated, most of the energy transfer goes on in water, because the water presents the largest number of targets for the radiation. Solute molecules, because they are *relatively* scarce, will be directly acted upon infrequently. Chemical changes, however, can be brought about in them if the energy of the ionizing particles is transferred to them from the ionized water. It is obvious, then, with respect to changes brought about in the molecules of which cellular constituents are composed, that the interaction of ionizing radiation and water, and the chemistry (the reactions possible) of irradiated water, will be very important. The degree of change brought about by radiation in the constituents of the cell depends upon that brought about in the molecules of which the cell is composed. The number of alterations, for example, in the structure or function of chromosomes, of mitochondria, or of any organelle depends upon how many molecules within the organelle have been changed. The changes in these molecules following irradiation are dependent primarily upon their interaction with the products of irradiated water and, to a lesser but by no means insignificant extent, upon the direct interaction of the molecules with radiation.

3.5 The Interaction of Ionizing Radiation and Water. When irradiated, water, like any other material, is ionized. An electron is removed from the molecule, leaving behind an ionized water molecule.

1. $H_2O \xrightarrow{\text{radiation}} H_2O^+ + e^-$ (electron)
2. $e^- + H_2O \longrightarrow H_2O^-$

Equation 1 depicts the ejection of an electron from an orbit of a water molecule. The molecule is now positively charged, and the electron which has been ejected is traveling with some discrete energy through the medium.

The reaction in Equation 2 typically follows that in Equation 1. An un-ionized or intact water molecule captures an electron (one perhaps set free in a reaction such as that in Equation 1). The result is another ion—a water molecule now endowed with a negative charge. A pair of ions (H_2O^+ and H_2O^-) has been formed.

The above reactions can, however, represent only a first step in a series of reactions because, when water is irradiated, it has been shown that the final products include H, OH, H_2O_2 and HO_2, none of which is formed as an *immediate* result of the passage of an ionizing particle through a water molecule. They must in some way have been formed from the two ions that are produced by irradiation (H_2O^+ and H_2O^-).

3.6 The Further Reactions of Ionized Water. The ions H_2O^+ and H_2O^- are not stable; they are not believed to persist in this form for more than a small fraction of a second before undergoing some change. They are said to dissociate almost immediately (10^{-16} seconds) into entities that are called *free radicals*.

3.7 Free Radicals. Free radicals are distinguished by the presence of a single, unpaired orbital electron, unpaired from the point of view of direction of spin on the electron's own axis. Electrons, in addition to moving about the nucleus of an atom in an orbit, also rotate upon their own axes; the direction of this rotation may, however, differ.

Within any orbit, electrons are paired with respect to spin so that for each electron spinning in one direction there will be another spinning in the opposite direction (this is known as Pauli's exclusion principle). If, for some reason, there are an odd number of electrons in an orbit, pairing cannot occur, and there will be *one electron* rotating upon its axis for which there is none spinning in opposition. An atom or molecule having such an unpaired electron is a free radical. Free radicals are exceedingly important in the study of radiation effects, for it is through them that the *indirect action* of radiation occurs.

3.8 The Formation of Free Radicals from Ionized Water. The outer electron orbits of H_2O^+ and H_2O^- may be represented in the following manner:

$$H_2O^+ = H\cdot\ \cdot\ddot{O}\cdot\ \cdot H$$

$$H_2O^- = H\cdot\ :\ddot{O}:\ \cdot H$$

There are seven and nine electrons respectively. These ions almost immediately dissociate into two subunits:

$$H_2O^+ \longrightarrow H^+ + OH\cdot$$
$$H_2O^- \longrightarrow H\cdot + OH^-$$

Each water ion will yield one smaller ion (H_2O^+ gives H^+ and H_2O^- gives OH^-) and one free radical (there is no way that either seven or nine electrons may be split between the two subunits so that both will have an even number). These are $H\cdot$ and $OH\cdot$ (the dot symbolizes the unpaired electron). The free radicals thus formed are responsible for the indirect action of radiation. They are extremely reactive; in pure water they ordinarily react within 10^{-5} second. If a solute is present their reaction will usually have been accomplished in even less time.

3.9 Reactions of the Free Radicals.

The energy exchanged into water from the ionizing particle ionizes the water. The water ions formed this way dissociate, yielding free radicals. The free radicals, then, will diffuse through the irradiated system, reacting with and producing chemical changes in anything with which they interact. In this way the energy of an ionizing particle is exchanged into water and, from water, to another, possibly an intact, molecule.

The reactions of free radicals are reasonably indiscriminate; a free radical may interact with another free radical, with a molecule already damaged by radiation, or, most important, with an intact molecule, possibly a solute molecule—one previously unchanged by radiation.

The reactions of the free radicals can be subdivided into five general categories.

A. Commonly, there will be a number of reactions among the free radicals themselves.

1. $H\cdot + OH\cdot \longrightarrow H_2O$
   Water is reconstituted.
2. $H\cdot + H\cdot \longrightarrow H_2$
   Molecular hydrogen is formed.
3. $OH\cdot + OH\cdot \longrightarrow H_2O_2$
   Hydrogen peroxide is formed.

B. Free radicals may react with the water in which they are formed:

$$H\cdot + H_2O \longrightarrow H_2O + OH\cdot$$

C. They may react with their own reaction products.

1. $H_2 + OH\cdot \longrightarrow H_2O + H\cdot$
   Molecular hydrogen is a reaction product in A.2 above.
2. $H_2O_2 + OH\cdot \longrightarrow HO_2\cdot + H_2O$
   Hydrogen peroxide is a reaction product in A.3 above.
3. $HO_2\cdot + HO_2\cdot \longrightarrow H_2O_2 + O_2$
4. $HO_2\cdot + OH\cdot \longrightarrow H_2O + O_2$
   $HO_2\cdot$ is a product of the reaction between $H\cdot$ and molecular oxygen, O_2.

D. Free radicals may react with oxygen. This reaction is of considerable importance and is treated in detail in Chapter 7. It will not be discussed here save to state that it does occur and that it enhances the effects of radiation.

E. Finally, and very important, free radicals may interact with organic molecules—the molecules of which cells and tissues are built—and change them. When this takes place the *indirect action* of radiation has occurred. The first step in the process is the exchange of energy in water (the solvent) from an ionizing particle. But the last step is a change in one of the molecules of *solute*, one of the molecules of which the cell is built.

The following equations illustrate how this may come about.

1. $HO_2\cdot + RH \longrightarrow R\cdot + H_2O_2$
2. $RH + HO_2\cdot \longrightarrow RO\cdot$
   (an organic free radical) + H_2O

If RH in either of these cases is a fundamental organic molecule—one important in the metabolism of the cell, either as a building block or as a finished product—an upset in the chemistry or the metabolism of the cell can be expected. In addition, H_2O_2 (hydrogen peroxide) is a cell poison and if present in sufficient quantities can materially interfere with metabolism.

3.10. Products of Interactions with Free Radicals.

Not all the products of free-radical interactions are harmful to living systems. Water is a product; so is molecular hydrogen. And, after a free radical has interacted, it is, itself, extinguished; it no longer exists. But this does not mean that all danger is past for the remaining cellular constituents. Some of the products of such reactions are poisons; still others are free radicals themselves, capable of further reaction and, consequently, further transfer of the energy of the ionizing particle. Organic free radicals may represent not only changed molecular constituents of the cell, but also substances that are free to attack other such constituents and spread molecular change still further.

3.11 Sources of Radiation-Produced Free Radicals.

It must not be imagined that free radicals will be formed only from irradiated water. The ionization of nearly any cellular component (fats, in particular) can result in free-radical formation which, in turn, will contribute to the indirect action of radiation. Water constitutes, however, the most abundant species of molecules in cells and, consequently, will play the largest role in this phenomenon. But, direct action upon any molecule which results in its ionization may also result in the formation of free radicals from the ion. Indirect action

primarily, but not exclusively, occurs from water-derived free radicals.

3.12 The Influence of Ion Density on Free-Radical Interactions.

The interactions of free radicals both among themselves and with their own reaction products is dependent primarily on how closely together the free radicals have been formed. After they are formed, they must diffuse through the medium in which they find themselves until they encounter something with which they may interact. Densely ionizing radiations (alpha particles, protons, electrons) produce clusters of ions that are very close together. These ions may subsequently dissociate into free radicals which also will be in close association with one another. Since free radicals can and do interact with each other, and since, after exposure to densely ionizing radiation and the dissociation of those ions, *they* are very close to each other, they need not diffuse far before encountering something with which they can interact. This something will be another free radical. Consequently, there will be a high probability of interactions between free radicals and of free radicals with the products of previous radical-radical interactions.

If ionization, on the other hand, is brought about by sparsely ionizing radiation, the clusters of ions formed are more widely separated than those produced by densely ionizing radiation. The probability of interaction between the resultant free radicals is much smaller than that following irradiation with densely ionizing particles. As the energy transfer from ionizing radiation within a given volume increases, there is an increase in free-radical interactions and an increase in the number of products of these reactions. The products are, in themselves, at least potentially harmful so that, as the density of ionization increases (increasing LET), there is likely to be an increase in the number of changed molecules—an increase of radiation effect—in the cell.

3.13 The Role of Oxygen.

The presence of oxygen during irradiation enhances the magnitude of radiation effects. These effects are of sufficient importance to merit special treatment (see Chapter 7). It is enough to state here that oxygen reacts with the free radicals produced by radiation and draws them into further destructive reactions with the molecules of the cell.

3.14 Influence of Hydration on Radiosensitivity.

If the formation of free radicals from ionized water is the major mechanism through which the indirect action of radiation proceeds, then the degree of hydration of a living system should influence its radiosensitivity. The more water there is, the greater the number of interactions with ionizing radiation that can go on in it, and the greater the number of free radicals that will be formed from it.

A solution in which water is the solvent may be considered more or less hydrated if it is more or less concentrated. The greater the concentration of solution, the dryer the solution may be said to be. More dilute solutions can be regarded as hydrated or wetter. By testing the effects of radiation on solute molecules in solutions of differing concentrations, an appreciation for the part played in changing these molecules by water-derived free radicals can be gained.

3.15 The Influence of Hydration on Radiosensitivity: Observations in Solutions. *The Dilution Effect.*

In solutions a fixed number of free radicals are produced by any given dose of radiation. Hypothetically, if the action of radiation in a solution, that is, the number of *solute* molecules chemically changed as a result of irradiation, were to occur only through the mechanism of indirect action (only by

free radicals), the number of these molecules actually changed would be *independent* of their concentration (except at extremely low concentrations). This is so because the number of free radicals which will bring about the changes from any given dose of radiation is limited. This number of free radicals can react only with a certain, fixed number of solute molecules. The availability of more or less *solute* molecules, then, will have no bearing on the total number of these molecules that are changed. If, for example, at a given dose of radiation, sufficient free radicals to change ten solute molecules are produced, it will make no difference whether ten, twenty, thirty, or more such molecules are present. Ten and only ten will actually be changed. If indirect action were the sole mechanism by which solute molecules in solution are changed, then the dose of radiation and the number of free radicals it produces will be the limiting factor and the only factor determining the number of solute molecules that actually are affected.

On the other hand, if, hypothetically, the action of radiation is exclusively direct (that is, direct ionization of solute *or* solvent molecules, but no transfer of energy from one to the other is possible), the number of *solute* molecules chemically changed will depend upon (in fact, is proportional to) their concentration. Since, in this instance, chemical change can be brought about only by direct ionization, the more solute molecules present, the greater the probability that some of them will be in the path of an ionizing particle and will be changed. The actual number "hit," then, will be proportional to concentration.

Of course, while it is true that both direct and indirect actions actually go on in any solution, observations made in more or less concentrated solutions can give a good idea of the importance of hydration and, consequently, the importance of *indirect* action. Such observations are made by irradiating solutions of different concentrations and comparing the *actual* results with those expected or predicted from the two hypothetical cases just described. Stated another way, if solutions of varying concentrations are irradiated with given doses of ionizing radiation, will the number of solute molecules changed depend upon the concentration of the solute molecules, or will the number be essentially independent of the concentration?

Experimental evidence has shown that, in dilute enzyme solutions of varying concentrations,[1,2] inactivation of enzyme (chemical change) produced by radiation is independent of concentration. Figure 3.1 shows schematically the results of a typical experiment. These give strong evidence that enzyme inactivation has come about primarily through the action of free radicals (indirect action) and that direct action upon the enzyme molecules has played a small, if not insignificant, role.

In extremely dilute solutions (those closest to zero in Figure 3.1) the relatively constant relationship between dose and inactivation no longer holds. The solutions are so dilute—there are so few solute molecules—that the free radicals which have been formed react with each other and not with enzyme. The free radicals interact and, in so doing, extinguish themselves, rather than inactivating (bring about change in) enzyme.

3.16 Observations in Living Systems. As a general rule, living systems containing little water are much less sensitive to radiation than organisms which exist in highly hydrated states.[3,4,5] It must be borne in mind that this is a generalization and that not every experiment comparing the radiosensitivity of the same organism in both the dry and wet states has shown unequivocally that radiosensitivity is greatest when irradiation is carried out in the hydrated condition. (Viruses; the seeds of higher plants; *Artemia*, the brine shrimp; and bacterial spores are examples of living systems in which such comparisons may be made.) However, desiccation and hydration are perhaps more complex biologic processes than the mere addition

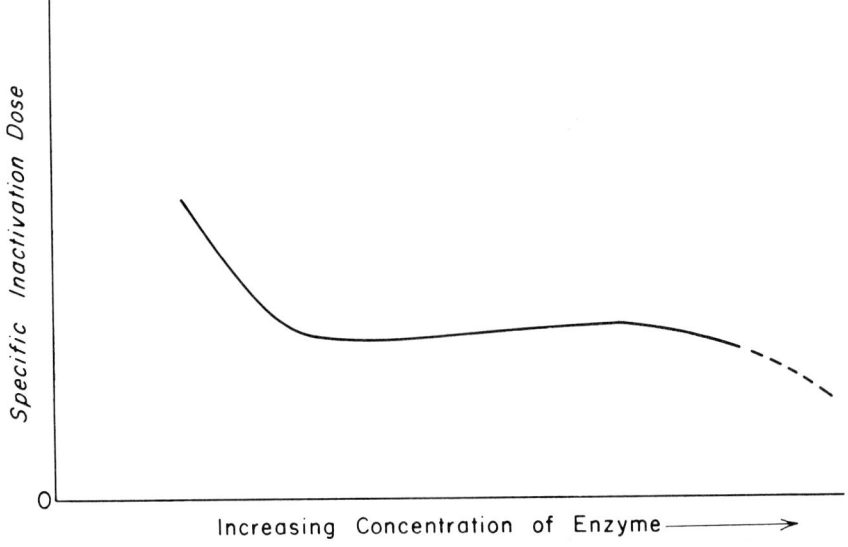

Figure 3.1. The relationship of dilution upon enzyme inactivation (chemical change) by ionizing radiation. Except for the most dilute solutions, there is a constant relationship between dose and inactivation which appears to be independent of concentration.

or subtraction of water would imply. Changes in rate or even in the mode of metabolism may accompany desiccation and hydration, and these might affect radiosensitivity. It is possible that such changes could mask or even reverse the expected influence of water upon radiosensitivity.

3.17 Proof of the Existence of Radiation-Produced Free Radicals in Biologic Systems.

While it is true that free radicals are formed subsequent to the ionization of water by radiation and while it is true that it is reasonable to assume that the interactions of radiation and cellular water would also produce free radicals, experimental demonstration of such free radicals in tissue (that is, those formed from *ionized· water*) after irradiation has not always been possible. At the usual metabolic temperatures, the life of water-derived free radicals is extremely short.

The length of time that free radicals exist as free radicals (their "life") is dependent primarily upon their rate of diffusion through the medium in which they are formed. They are formed, and then they move (diffuse) until they encounter something with which they interact. After interacting, they, as entities, become extinct, for they will now exist in combination with whatever underwent the reaction with them. (This is not to say that the product will not be another free radical. It means only that the interaction of a given free radical brings about its own extinction. At the same time it may give rise to something new.) The rate of diffusion, then, determines how long a time will elapse between free-radical formation and an encounter which results in reaction and extinction. Efforts to demonstrate the existence of free radicals in living things have been hamstrung by the exceedingly brief period between the time of the formation of the free radicals and that of their extinction as a species.

It can be supposed that indirect action will be inhibited in living systems irradiated in the frozen state (bacterial spores can be used in such instances) and that this inhibition can be taken as evidence for radiation effects being brought about by free radicals. These suppositions are made because the diffusion of free radicals through solid water will be greatly impeded. The impedance should make the degree of change at any given time following radiation less than that observed in nonfrozen systems. Methods and the results of such experiments are not directly applicable nor, unfortunately, extrapolatable to living systems of high normal body temperatures.

3.18 Significance of Changes in the Molecules of Which Cells Are Composed.

Any change brought about by ionizing radiation in the molecules of which cells are composed is, potentially at least, detrimental to the viability of the cell. It can be said that, because radiation brings about chemical changes, *any* exposure in *any* amount will be a potential hazard to living things. This is true because living things and the material of which they are composed, through millennia of selection and development, are in precise, harmonious balance with their various environments. Each molecule in every cell is part of a delicate system of checks and balances with every other molecule. Similarly, the cells in each tissue are in balance with each other; the tissues within the body function in coordination with each other, and organisms themselves are balanced with each other and with the physical environment. Within the cell, the balance between molecules is so finely adjusted that deviations from it produce flaws in essential cellular activity and work to the detriment of the cell as a whole. If the status quo within cells is changed, serious imbalances or errors in metabolism and function, even some that are lethal, can be the result.

Ionizing radiation, by depositing energy in cells through the ejection of electrons, upsets this status quo—this finely attuned balance. The total amount of energy in the cell is changed. But, more than that, the products of irradiation—the products of interaction between ionizing particles and the atom with which they chance to interact (ions, excited atoms or molecules, and free radicals)—may be of a kind foreign to the cell or, under normal circumstances, of a kind present only in smaller quantities. Reactions may go on between these radiation products and the remaining normal cellular constituents that usually do not or cannot occur. Thus, because any dosage of ionizing radiation, regardless of how small, has some probability of exchanging some of its energy within any cell which it traverses, there is no dose of radiation too small to be without any hazard at all.

3.19   The Relative Importance of Direct and Indirect Action in Biologic Systems. In a certain sense it is meaningless to speak of the relative importance of direct and indirect action. Whether molecules are changed by direct interaction with radiation or by interaction with the products of an exchange of energy with another molecule will be of little importance to a biologic system. A changed molecule—a deviation from the status quo—is the result. Yet, in another sense the two modes of action will have a relative importance, for, although it is true that any energy exchanged to the molecules of the cell is potentially dangerous to the cell, changes brought about in certain molecules will be more certain to damage cells than others. That is, not every molecule is of equal importance to the cell although every *species* or kind of molecule is indispensable. Some species of molecules of which living things are composed are, themselves,

made up of a very large number of individual molecules. Often, more are present at any given time than are necessary or essential. Although these are present within cells, each and every one of them will not enter into the cell's metabolism or become part of the structure of the cell. Enzymes are a good example of such a species; water itself is another example. Both are present in an excess. Changes brought about in certain kinds of enzyme molecules may do no harm to the cell at all if the changed molecule is one of those that never enters into metabolism. If that particular molecule never functions, that particular radiation effect is disarmed. On the other hand, one can never be certain which of the large, even excessive, number of enzyme molecules will become functional. At present, there is no way to tell. Whenever an enzyme molecule is changed, then, the change is *potentially* harmful to the cell. This is true because any enzyme molecule has the *potential* to become functional. If the radiation-changed enzyme molecule is one which is called upon to function, the change in its structure may very well have brought about a change in its function or impaired its ability to function in the cellular situation in which it is needed. The metabolism of this cell will be, for that enzyme and for that function, impaired. There will have been a deviation from the status quo; the system of intramolecular checks and balances will have been upset. The cell may or may not be able to overcome this difficulty and restore the proper balance. If the cell can overcome it, the radiation effect will have been temporary and it may be thought of as reversible, possibly even as "repairable." If the cell cannot, the cell will be permanently damaged; within the cell the molecular environment will be out of balance. Further, the cell itself will now be out of balance with the others in the tissue of which it is a part.

Other species of molecules, on the other hand, are composed of only a few members. The members are not present in excess, and each of them takes an active, often critical role in metabolism. The nucleic acids exemplify this type. Changes brought about in nucleic acid molecules, in particular in deoxyribonucleic acid molecules (all metabolism is under the direction of these molecules), will almost certainly bring about a defect or derangement of normal metabolism because each of the molecules is intimately involved in cellular, even intracellular, activity. (See Chapter 9 for a discussion of the role of nucleic acids.) Each of the molecules of deoxyribonucleic acid may be said to constitute a critical or vital target for radiation. At the same time, however, the number of such molecules is small. Therefore, the likelihood of *direct* interaction of *randomly* interacting radiation with one of these molecules is also small. By direct action of radiation, changes brought about in these critical or *key* molecules are rare. It is in this case that the indirect action of radiation assumes its greatest importance. The irradiation products of more common but less vital molecules can transfer energy to and bring about changes in one of these key molecules. By indirect action, then, changes in the vital, key molecules become rather more common, and serious, usually irreversible, deviations from normal metabolism are brought about. From the point of view of changes in key molecules, from the point of view of serious deviations from the status quo, indirect action is more important than direct action.

3.20   Protective Agents.   Cells appear to have no way to protect themselves against direct action, which, after all, results from radiations entering cells in an unpredictable manner from the environment. Cells are equipped, however, with means to ameliorate the potential dangers from in-direct action. Compounds with an affinity for reaction with free radicals are present in most cells so that, when free radicals are suddenly formed, there is a high probability they will react with such molecules rather than with important organic molecules. Sulfhydryl proteins (R–SH) are compounds commonly found in cells that serve this purpose.

SUMMARY

1. The action of ionizing radiation (excitation, ionization) is random; any atom with which it interacts is likely to be ionized.
2. In mixtures the substance present in the greatest amount has the greatest probability of interaction with radiation; those present in lesser quantities are involved less often.
3. Ions produced during irradiation can dissociate into free radicals.
4. Free radicals can interact with and bring about changes in molecules not themselves directly changed by radiation. This is called the *indirect* action of radiation. It is the process by which energy from ionizing radiation is transferred from a directly ionized molecule to one not directly ionized.
5. No amount of radiation is too small to produce no potential deviation from normal cellular metabolism.
6. Indirect action is the most important mode of change of key cellular molecules and in the production of sharp deviations from the cellular status quo.

Text References

1. Dale, W. M., Gray, L. H., and Meredith, W.: Phil. Trans. A., *242*:33, 1949.
2. Dale, W. M.: The effect of x-rays on the conjugated protein d-amino-acid oxidase. Biochem. J., *36*:80–85, 1942.
3. Hollaender, A.: Symposium on Radiobiology. New York, John Wiley, 1952, p. 285.
4. Hutchinson, A.: Radiat. Res., *7*:468, 1957.
5. Engel, D. W., and Fluke, D. J.: Radiat. Res., *16*:173–181, 1962.

Chapter 4

INTRACELLULAR RESPONSES:
TARGETS

4.1 Introduction. The target theory concerns itself with possible small sites or regions in cells that are believed to control various cellular functions. If sufficient radiation energy is absorbed in them, they may be changed and become inactive so that the function or functions either carried out by or controlled by these sites or targets are lost. The theory rests on the idea that if all parts of cells are equally sensitive to radiation, then all identical cells absorbing the same dose of radiation should be affected to the same degree. But this is not the case. Equal amounts of radiation absorbed in identical cells do not always produce the same effects or degree of effect. Densely ionizing, high LET radiations are most effective at damaging cells, and it may be because they have a high probability of exchanging large amounts of energy into small volumes such as might be occupied by these targets. Radiations having low LET spread their energy deposition out more or less uniformly in cells, so that no portion of given cells receives large amounts of energy (Figure 4.1).

4.2 Survival Data. Survival data collected on cells grown *in vitro* and ir-

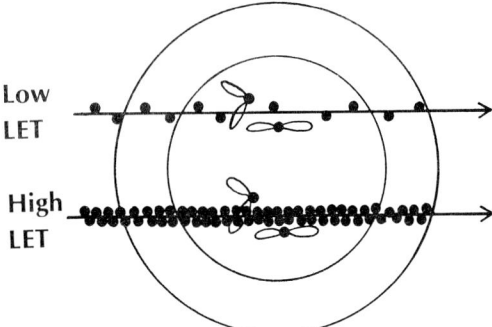

Figure 4.1. Diagram of a cell irradiated with equal doses of radiations having high and low LET. The targets or sensitive sites in this illustration are the chromosomes, structures in the nucleus. The low LET radiations pass through chromosomes but exchange little energy in any of them. Although the high LET radiation has a lesser probability of passing through any given chromosome, when it does, it exchanges much energy into it.

radiated with x rays tend to support this idea. When identical cells are cultured and irradiated using differing doses of x rays and the survival of these cells is observed, two typical responses are observed (Figure 4.2). These are exponential and sigmoidal in nature.

The end point "survival" measured here is not life or death, but survival of the capacity to reproduce. Reasons for using

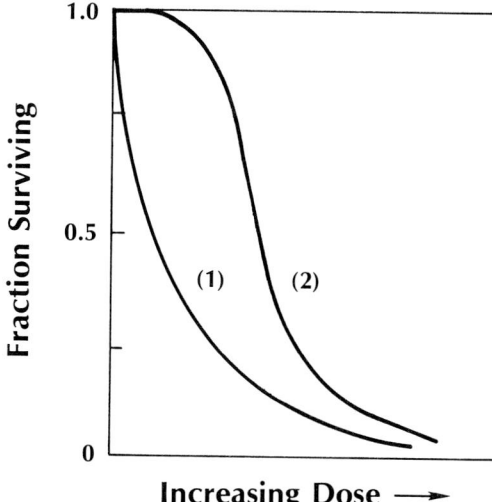

Figure 4.2.   Typical survival curves resulting from x-irradiation of identical cells grown in culture. Curve (1) is an exponential expression and curve (2) a sigmoidal expression.

"zero" and there is no deviation from the exponential relation over a very wide range of dose. This suggests that previous absorbed radiation does not affect cell killing. The same proportion of cells is killed by any given quantity of radiation whether those cells have absorbed radiation or not. The dose-survival relationship is all-or-nothing. Either cells sustain a lethal injury and die or they do not, and it does not matter how much previous irradiation they have had. There is no suggestion of accumulation of radiation damage until it reaches a level where reproductive capacity is lost, no suggestion that damage at a sublethal level to the reproductive system is occurring. The relationship further suggests that if a radiation interaction producing a single critical change occurs in a single critical region or target in the cell, the reproductive system fails. If that critical interaction does not occur and the change is not made, the reproductive system operates, no matter how many other interactions have occurred in that same cell. The data do suggest the existence of a single target in this kind of cell and the interaction of radiation or its by-products with this target to change it before the ability to reproduce is lost. Even when such cells absorb radiation energy, if the energy is not absorbed in the target or if the target is unchanged in a specific way, they may still reproduce.

Such observations have led to the following hypothesis: that cells obeying an exponential dose-survival response have one critical target governing reproductive capacity, and that a single critical change must be made in that target to inactivate it. The hypothesis is known as the "single hit:single target model."

this end point are: (1) In the way the experiments are done, colonies of cells derived from individual survivors are scored. Cells that live through irradiation but that cannot produce a colony of descendants (are sterile) are not included. (2) Biologically the inability to produce a line of descendants, the inability to reproduce is, in a sense, equal to extinction. Sterile individuals cannot pass on their genotypes and can contribute nothing to the history of a species.

4.3 Exponential Curves. In an exponential dose-survival relationship (see Figure 4.2, curve (1)), any given quantity of radiation dose kills the same proportion of cells as every other like quantity, no matter where that dose is along the X or dose axis. For example, the proportion of the total number of living cells killed by 100 rads will be the same whether those hundred rads are the first hundred rads given the culture (between 0 and 100 rads) or whether it follows considerable previous irradiation (say, between 700 and 800 rads).

Cell killing begins immediately after

4.4 Sigmoidal Curves. Cells that display a sigmoid dose-survival response are different. Sigmoidal curves can be resolved into three components: a slight negative slope from the origin through very low dose ranges, an inflection called a shoulder, and an exponential survival response through the remaining, higher dose range. Figure 4.3, which follows, shows both an exponential and a sigmoid relationship plotted on coordinates of which one is a logarithm. When such a coordinate system is used, exponential relationships appear as straight lines. It is important to stress that these relationships, which appear as straight lines on semilogarithmic coordinates, do not depict linear relationships but are still exponential. This is a more useful way to display survival data than the familiar curved line and will be used throughout the remainder of the text.

If an exponential dose-survival relationship is understood as the inactivation of a single critical target (n = 1), then through the *exponential* region of the sigmoid curve it must be assumed that cells have only one target. On the average, cells surviving that radiation dose above which dose-response is *exponential* (see Figure 4.3, curve 2, region C) are believed to have one remaining target. But some quantity of radiation dose, which varies somewhat from cell population to cell population, had to be absorbed by the population before the target number in surviving cells was brought to one (see Figure 4.3, curve 2, regions A and B).

Through this lower dose range (the region of negative slope and the "shoulder"), the dose-survival relationship is *not* exponential, and the same proportion of cells is not killed by each like radiation dose. In fact, the number killed is *higher* at the higher end of the dose range than at its lower end. For example, 10 rads of radiation kill a smaller proportion of cells in region A near zero dose, than does 10 rads of radiation in region B near the start of C. Since all cells receive and presumably absorb the same quantity of radiation energy, it means this energy is not equally effective at producing the end point (loss of capacity to reproduce) through the low dose range. Because the radiation be-

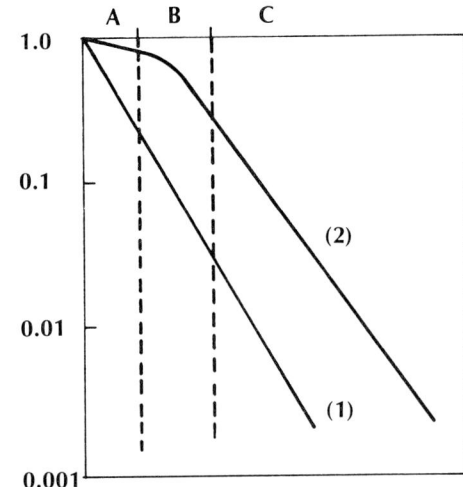

Figure 4.3. An exponential (curve 1) and sigmoid (curve 2) relationship displayed on semilogarithmic coordinates. The dose is on a linear coordinate, and the surviving fraction is on a logarithmic coordinate. The exponential relationship is a straight line, but the sigmoid relationship has a region of slight negative slope (A), a shoulder region of steeply changing slope (B), and an exponential relationship (C).

comes more *effective* as dose is increased through this dose region (A and B), this relationship suggests that damage from radiation is accumulating, piling up in cells of the population until a critical or threshold quantity is reached. At that threshold the relationship of dose to survival becomes exponential, and there seems to be only one further remaining increment of damage possible (one target only) before the reproductive function is lost.

In cells displaying sigmoidal dose-survival relationships, then, failure of the reproductive system does not seem an all-or-nothing phenomenon; on the contrary, it seems that damage accumulates and that only after sufficient accumulation is there loss of reproductive capacity. Damage to this function does not always result in loss of reproductive ability, and such cells do not conform to the single hit: single target model.

A sigmoidal dose-response relationship, on the other hand, suggests that more than a single critical change must be made, more than one unit of damage *must* be caused before reproductive capacity fails. Cells displaying a sigmoidal dose-survival relationship seem to have more than one target, more than one sensitive site, and all such targets must be changed and inactivated before cell reproductive capacity is lost. While it is true that cells displaying sigmoid kinetics seem to have a number of targets, it is still believed that a single critical event is all that is required to inactivate any one of them. This notion is derived from the smooth, uninflected nature of the exponential region of the survival curves. If some of the many targets in multitargeted systems were different from each other in this regard, changes in slope of these exponential regions would be expected. But, except for populations that are known to be heterogeneous, this does not occur. Does this mean that all sensitive sites or targets controlling reproduction in multitargeted cells are the same things? One cannot be sure. The fact that single critical events inactivate any of them merely means they respond in the same way to radiation, but this cannot be said to mean they are all the same.

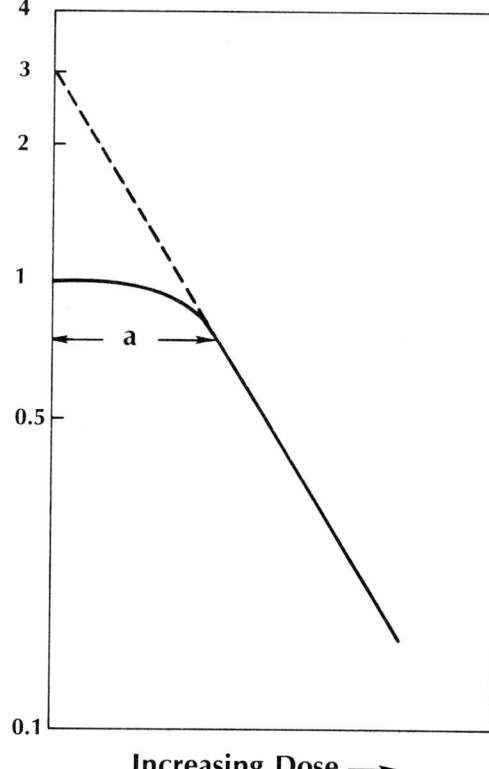

Figure 4.4. Extrapolation of the exponential of sigmoid survival curves to intercept the Y ordinate gives the *average* number of targets per cell in the population. The width of the shoulder is measured at a.

4.5 The Extrapolation Number. It is possible to make estimates of the number of targets in cells of populations having more than one target. This is done by extrapolating the exponential segment of sigmoidal dose-survival curves to intercept the ordinate (Figure 4.4). In the exponential portion of survival curves, there is only one target per cell and the probability of its being inactivated by any given quantity of radiation dose is given by the *slope* of the survival curve. Since all targets in given cells or cell populations seem to respond in the same way to irradiation, it is reasonable to assume that in the shoulder region of these curves (Figure 4.4, a), where there is more than

one target, the probability of targets being inactivated by given quantities of radiation dose is the same as for the single remaining target in the exponential region of the curve. Thus, the curve for *inactivation* of *targets* (not for survival of cell proliferative capacity) would be exponential throughout and have the same slope as the exponential region of the sigmoid curve (Figure 4.4). Extrapolation of the exponential region of sigmoid curves gives the relationship between dose and target inactivation. The point at which this extrapolated line intercepts the survival axis gives the average number of targets per cell at zero dose—when no radiation has been given. It is called the extrapolation number and is represented by the symbol n. This number is an estimate of the average number of targets in cells of given population and tells how much sublethal damage such cells can absorb. Although the extrapolation number is variable for cells of various populations, for many kinds of mammalian cells grown in tissue culture, it appears to be 2 or nearly 2. The extrapolation number may vary in given cells under a variety of circumstances, including changes in either the internal milieu of the cell or environmental conditions. These will be considered in some detail in later chapters.

The extrapolation number, because it is an average, does not need to be a whole number, and determinations of extrapolation numbers for various populations often turn up numbers between two whole numbers. Nevertheless, it is still true that extrapolation numbers in tissue-cultured mammalian cells often are seen to be near 2.

SUMMARY

1. Cells are believed to contain small regions called targets, which are re-

sponsible for controlling various functions.
2. Certain cell types have a single target controlling reproduction, and a single critical event can inactivate both target and the reproductive mechanism.
3. Certain cell types have more than one target controlling reproduction. A single critical event inactivates any one of the targets, but all must be inactivated before reproductive capacity is lost.
4. The averge number of targets in cells of given populations is derived by the extrapolation of the exponential portion of sigmoid survival curves to the Y intercept of such curves.

General References

Elkind, M. M., and Whitmore, G. F.: The Radiobiology of Cultured Mammalian Cells. New York, Gordon and Breach, 1967.
Lea, D. E.: Actions of Radiations on Living Cells. New York, Cambridge University Press, 1956.
Puck, T. T., and Marcus, P. I.: Actions of X rays on Mammalian Cells. J. Exp. Med., *103*:653, 1956.

Chapter 5

INTRACELLULAR RESPONSES:
RADIATION SENSITIVITY

5.1 Introduction. Radiation sensitivity may be expressed as the amount of radiation energy required to produce a given end point or a given degree of severity of radiation-produced change. In this chapter the term will be used in connection with the amount of radiation necessary to destroy targets governing cellular reproductive capacity.

The sensitivity of targets governing cellular reproductive capacity to radiation varies among different cell types and in a number of circumstances. It can be determined from survival curves. The *slope* of the *exponential* portion of survival curves gives it. Curves having steep slopes are derived from radiosensitive populations, and those less steep are from more resistant cell populations (Figure 5.1).

Cell populations that have the exponential portions of their survival curves parallel, have targets with the same radiosensitivity; those in which the exponentials are not parallel differ in radiosensitivity.

5.2 Radiosensitivity and Extrapolation Numbers. As can be seen from Figure 5.1, radiosensitivity and extrapolation numbers are not related. Extrapolation numbers represent the amount of damage

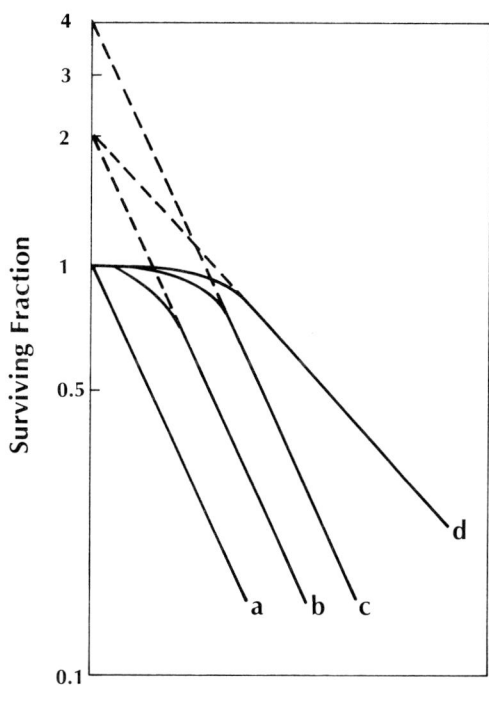

Figure 5.1. Of the four cell populations represented by these curves, targets governing the reproductive mechanism of populations a, b, and c have the same radiosensitivity, but d is different. The slope of the exponential portion of these curves gives the radiosensitivity, and populations having parallel exponentials (or the same slope) are equally radiosensitive. Populations a and c differ in their extrapolation numbers, but those of b and d are the same.

35

that can be absorbed by a cell population (the number of targets); radiosensitivity is the amount of radiation energy required to inactivate each target. In Figure 5.1, the targets are equally sensitive to radiation in populations a, b, and c but have a different sensitivity in d. There are *more* targets, greater degrees of possible sublethal damage, in c than in b and in b than in a, but the amount of radiation energy needed to inactivate any of these targets (cause irreparable damage to any of them) is the same. Populations d and b may absorb the same number of units of damage (same n), but the amount of radiation energy needed to produce a unit of damage is less in b than in d. The targets of d are, for some reason, more resistant than those of b, and this comparative difference is shown in the slopes of their curves.

The sensitivity of cellular reproductive capacity to radiation, is, as is the case with other functions, not constant. It varies and does so according to variations in cellular activities (such as a phase of cell life cycle), as well as in response to extracellular, environmental factors. The same is true of n, the extrapolation number. The amount of sublethal damage that can be tolerated varies and, again, according to intrinsic cellular and extrinsic environmental factors.

5.3 D_0 and D_{37}. It is also possible to derive a measure of radiosensitivity that can be expressed in units of absorbed radiation, rads. Since radiation sensitivity may be defined as the amount of radiation energy needed to irradiate each of the targets in cells, a consideration of *pattern* of target inactivation is in order. First, as stated in Chapter 4, all targets in given cells are believed to respond the same way to irradiation and, from that point of view, are indistinguishable from one another. Although cells possessing more than one target lose their reproductive capacity according to a sigmoid pattern,

the *targets in* these cells are inactivated according to an *exponential* pattern. But why is this so? If any and all targets require only one "hit," one critical radiation-related change for inactivation, why doesn't target inactivation proceed according to a linear relationship?

This would, in fact, be true if radiation-target interactions were selective and if radiations "hit" only intact, still functioning targets. But radiation-matter interactions are, instead, random, and inactivated targets carry exactly the same risk of another radiation interaction (another hit) as intact targets. When irradiation of a cell population is begun (when dose is "zero"), all the targets in the population are intact, and nearly every radiation-target interaction has a chance to inactivate a target. When most targets are intact, inactivation *is* linear through very low dose ranges. As a dose is increased, however, the number of inactivated targets begins to grow and forms a significant proportion of the total. A significant portion of the radiation will therefore exchange energy in targets *no longer capable of responding,* those which are already inactivated. When that point on the dose scale is reached, a deviation from linear relation is expected; of any given quantity of radiation used, fewer intact targets will be inactivated than were previously the case. As dose is further increased and the *proportion* of targets (of all those present) that are inactivated grows, more and more of any given radiation quantity is wasted interacting in them, and fewer intact targets are "hit" (Figure 5.2).

Although large amounts of radiation energy are required to inactivate all targets, the amount effective (not wasted) in target inactivation can be arrived at by extrapolating the slope of the initial linear relationship to the X axis (Figure 5.2). This point of intercept is called D_0. Radiation energy in amounts greater than D_0 is, in a sense, wasted, hitting targets now unre-

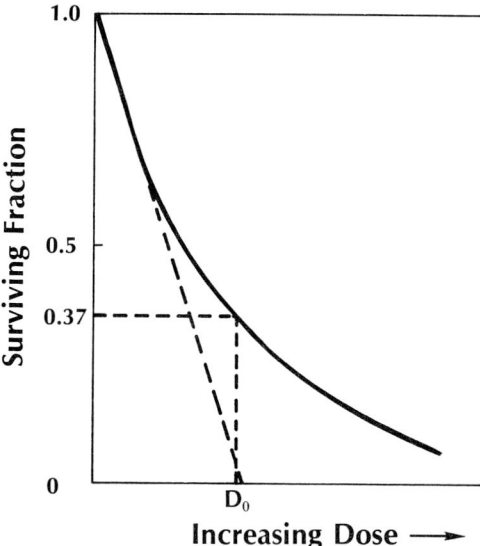

Figure 5.2.   Target inactivation as a function of dose. Through low dose ranges target inactivation is linear, but a deviation, resulting in an exponential relationship, occurs as dose increases. If the linear relationship were to continue throughout, all targets would have been inactivated at D_0, the point at which this extrapolated relationship intersects the abscissa. Because of radiation wasted in already inactivated targets, all targets are not actually inactivated until much higher doses are reached.

sponsive, already inactivated. D_0 is a measure of radiation sensitivity because it tells how much energy is *actually effective* in inactivating the targets in cells of any population.

Comparison of D_0 among various cell populations yields their comparative radiosensitivities. If the D_0 radiation dose for any population of cells is actually delivered to that population, all the targets will not, of course, have been "hit." How many targets will actually have been inactivated and how many remain intact? When D_0 has been given, 37 percent of targets will still be intact and 63 percent will have been inactivated. This will always be true, as it is inherent in an exponential relationship (see Figure 5.2).

Another measure or index of radiosensitivity then is D_{37}, the dose which leaves

37 percent of targets surviving or which inactivates 63 percent of them. Obviously, for targets that are inactivated exponentially or for cells in which there is only one target, D_0 and D_{37} are equal if the slopes of these curves are parallel. But if one applies these principles to *cells* with more than one target, there must be a modification. When various cell populations have shoulders on their survival curves of unequal widths, D_0 and D_{37} will not be equal *on the portion of the curve that includes the shoulder,* even if the exponential portions of the relationships are parallel. Once the shoulder is passed and the curves are on the exponential, D_0 and D_{37} will be equal because the dose required to reduce a population to any fixed proportion of the starting population depends on the width of the shoulder (Figure 5.3).

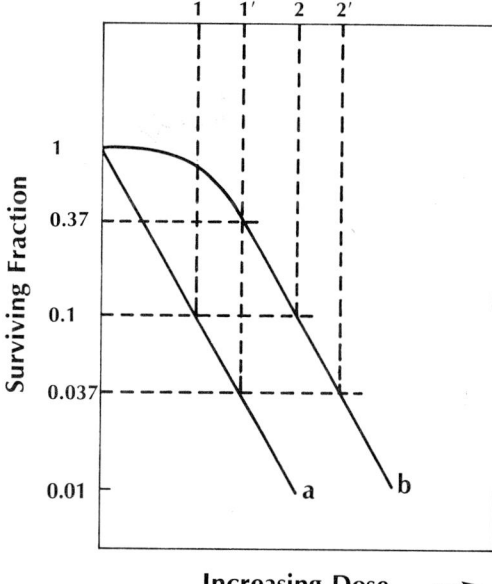

Figure 5.3.   The dose required to bring the populations represented by curves a and b to 0.37 (37 percent) survivors is greater in b than in a. This is related to the width of the shoulders. D_0 and D_{37}, however, are also given by the dose between 1 and 1', 2 and 2'. These doses are equal for populations a and b. D_0 and D_{37} is the same for populations a and b once the shoulder is passed, but different in the first survival decade.

5.4   Cell Killing and Radiation Sensitivity. The response of cells to irradiation, namely, loss of reproductive capacity, will be determined by the radiosensitivity (the energy required to produce a critical, inactivating change in the targets of those cells), as well as by the number of targets (the capacity for cells to absorb sublethal damage). Both of these may vary among various cell types and even among similar cells at various times in their lives or under a variety of environmental conditions.

It is a mistake to equate low target number with ease of cell killing, and it is also wrong to expect cells with large numbers of targets to be difficult to kill. Bacteria, typically, have a single target (n = 1), but it and therefore their reproductive mechanism is quite resistant to irradiation. Mammalian cells have rather sensitive targets, but owing to the large numbers of them that must be lost before sterilization occurs, reproductive integrity in these cells, too, is not easily destroyed by irradiation. The amount of radiation (the sensitivity) needed to make the single critical change in the target of bacteria is high, whereas that needed for mammalian cells is relatively much lower. *By both means*, either a single, resistant target or a large number of sensitive ones, the continuation of reproductive integrity in the face of radiation, at least at natural background levels, seems rather well assured.

5.5   Cell Killing. Cell killing by radiation, then, may be determined by either or both of these parameters, the radiation sensitivity and the number of targets. In cells with only one parameter, that parameter determines cell killing. But in those having more than one, the number of such targets and the sensitivity of each of them will be the determinants.

General References

Andrews, J. R.: The Radiobiology of Human Cancer Radiotherapy. Philadelphia, W. B. Saunders Co., 1968, pp. 25–33.

Elkind, M. M., and Whitmore, G. F.: The Radiobiology of Cultured Mammalian Cells. New York, Gordon and Breach, 1967.

Hall, E. J.: Radiobiology for the Radiologist. New York, Harper & Row, 1973, pp. 15–48.

Lea, D. E.: Actions of Radiation on Living Cells. New York, Cambridge University Press, 1956.

Puck, T. T., and Marcus, P. I.: Actions of X-rays on mammalian cells. J. Exp. Med., *103*:653, 1956.

Chapter 6

INTRACELLULAR RESPONSES: RECOVERY

6.1 Recovery. In cell types where sublethal damage to the reproductive mechanism is possible (cells conforming to a multitarget model), it is reasonable to expect at least the possibility of repair and recovery from radiation exposure. To determine whether this actually occurs, damage to reproductive capacity from a given radiation dose is determined after the dose is delivered, either at once or split in two or more fractions. When this is done to cell populations *in which the target number is greater than one*, less damage to reproductive capacity is apparent after exposure to radiation doses given in fractions than from the same doses given in single exposures. This suggests that unless multitargeted cells are killed by radiation they are able to recover from some or all the damage they have accumulated during radiation exposure. Radiation given in fractional exposures to a total accumulated dose produces less net damage to reproductive capacity than the same dose given all at once. Figure 6.1 is a typical survival curve obtained after a given radiation dose is administered in two fractions.

Such data clearly indicate that if cell types able to accumulate sublethal dam-

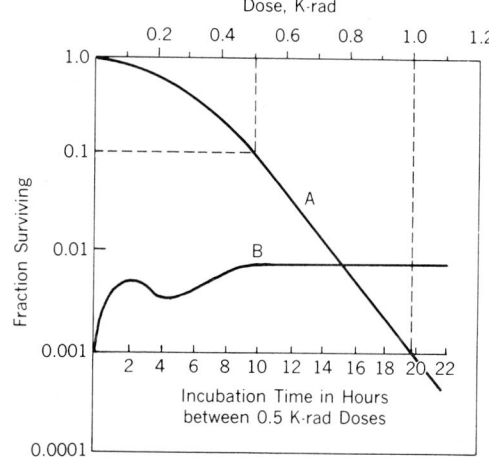

Figure 6.1. A hypothetical example in which the time lapse between radiation dose fractions is varied, but the size of the two fractions is equal. Curve A is the survival curve when no time elapses between x-ray doses. At 1.0 K-rad the fraction in which reproduction capacity survives is 0.001. The two fractions, 0.5 and 0.5 K-rads, given with no time between also reduces the survivors to 0.001. Individually each fraction reduces the population to 0.1. Curve B shows the fraction surviving when the first and second fractions are separated by incubation times given on the abscissa. When no time elapses between fractions, there is no change in the surviving fraction (0.001), but as time elapsing increases, changes are observed. Up to 2 hours in this example, increasing numbers of cells survive. This period is followed by an oscillation in which there is an increase in sensitivity, and this is followed by another increase in resistance and a long plateau.

age survive radiation exposure, they can recover from some or all damage done during that exposure. Given enough time, recovery (as represented by the number of cells surviving *both* fractions of radiation dose) reaches a maximum and no longer changes (the plateau of Figure 6.1). If cells that have recovered from radiation are again irradiated, they will again be able to absorb sublethal damage. How *much* sublethal damage they can absorb can be demonstrated by experiments such as the following. Populations of cells may be exposed to a series of radiation doses sufficient to reduce the target number to an average of one per cell. That is, radiation doses must be given such that cells surviving them will be on the exponential portion of the curve. This means, of course, that in these survivors the maximum sublethal damage has been done. Further radiation exposure has a high probability of inactivating their last remaining intact target, causing loss of reproductive capacity. Then, time is allowed to pass, sufficient time so that all recovery that is expected to take place will have taken place, and no further changes in recovery are expected. A time, then, on the plateau region of the recovery curve in Figure 6.1 would be chosen. After that time has elapsed, a second series of radiation exposures, the *same* series of doses as given the first time, is delivered to these survivors. The number of cells surviving the second set of doses is obtained, and a second survival curve constructed (Figure 6.2).

If experiments such as these are carried out, the *second* survival curve again has a shoulder. This means, as already demonstrated in the previously described experiment, that some recovery must have taken place; if there had been no recovery, the second survival curve (Figure 6.2, B) would simply have been an extension of the exponential portion of the first curve (Figure 6.2, A). However, in the time lapse between the two radiation

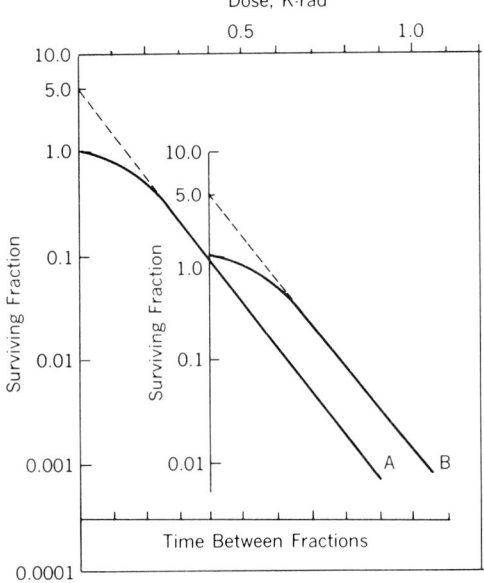

Figure 6.2.   Typical results of split-dose irradiation of mammalian cells when enough time for full recovery is allowed between fractions. The first, or conditioning dose, is constant, but the second is varied. Time between fractions is constant. Curve A is the split-dose response when no time elapses between fractions. Extrapolation of the exponential reveals n of about 5. Curve B results from split-dose exposures when time elapses between fractions. Extrapolation of the exponential reveals n of about 5.

exposures, the lapse in which recovery is known to take place, the target number is changed from one to a larger number. Further, as can be seen from Figure 6.2, the number of targets in fact returns to the *pre-irradiation* level. This suggests that unless all those sites governing the function of the reproductive mechanism in multitargeted cells are destroyed, they will in time fully recover and the cells' capacity to absorb radiation damage returns to that of unirradiated cells of the same kind.

Certainly such data support the contention that damage to the reproductive mechanism consists of inactivation of sensitive sites or targets in cells and that recovery from this damage consists of the repair or replacement (or both) of those targets. Full recovery seems to require

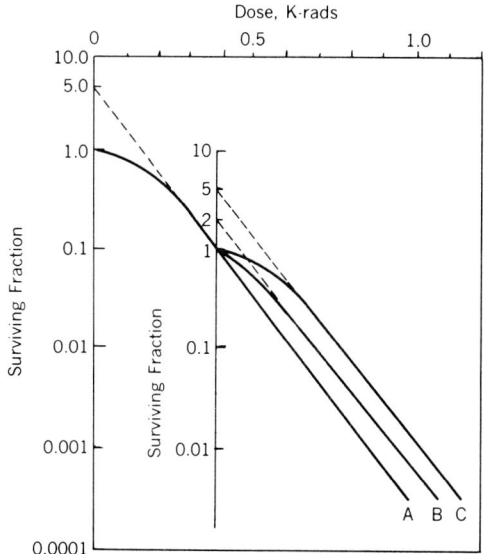

Figure 6.3. An illustration of the reappearance of targets with time after an initial dose of radiation. Curve A is the single-exposure survival curve. Curves B and C are split-dose curves, the second fraction of radiation given at different times after irradiation, with C later than B. Both Curves B and C lie above B. Extrapolation of the exponential of both curves reveals different values of n. n of Curve C is the same as that of Curve A. Such results are believed to mean that targets damaged by the conditioning fraction are repaired and represented as an increase in n with time. By the time the data for Curve C were obtained, recovery was apparently complete and the value of n is the same as in Curve A, the pre-irradiation value.

time, but if enough time is allowed to elapse between radiation exposures, full recovery is likely to occur. Irradiation before sufficient time for full recovery has been allowed results in only partial recovery and the appearance of smaller shoulders (Figure 6.3).

6.2 Cell Life Cycle.
In Figure 6.1 it can be seen that the effects of a fractional radiation exposure are not always the same. Even after cells reach their recovery peak, a second exposure sometimes shows a dip or oscillation in which the number of survivors from the total dose falls below that peak. The explanation commonly accepted for this oscillation is

that cells surviving the initial fraction pass, in time, through a radiation sensitive period. During this period their targets are more easily damaged by lesser quantities of radiation. The changing radiation sensitivity is believed to occur as cells progress through the phases of their life cycles.

If radiation dose-cell survival data are collected from cell populations in which there are cells in every phase of the reproductive cycle (Figure 6.4), it can be seen that not all phases of cell life are equally sensitive to radiation.

When cells are irradiated in most of the phases of the reproductive cycle, there is a delay in their progression from the phase

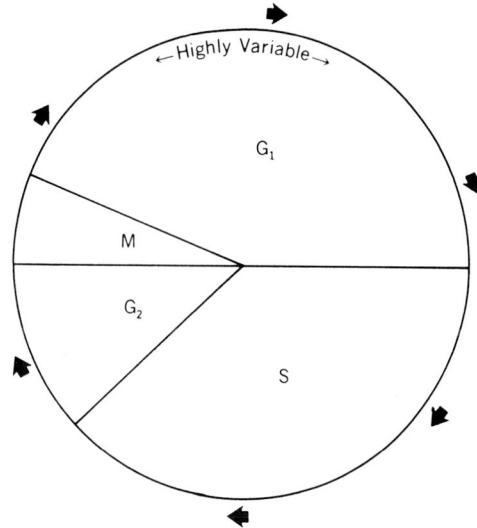

Figure 6.4. A diagrammatic representation of cellular life cycle. It begins after M, mitosis, so the earliest stage in G_1 is highly variable, probably ranging from a few minutes to many months, depending on cell type. Cells are thought of as "aging" as they progress through the cycle. S, the stage in which DNA is synthesized, is usually long compared to either G_1 or G_2, and M is usually very brief. These approximate proportions are typical of many mammalian cells, but individual types may deviate. The symbol G stands for the word *gap* because the processes going on in G_1 and G_2 are not completely known and are gaps in knowledge. Of these phases, the latter portion of G_1 and the first third of S, G_2, and M are sensitive. The remainder is resistant.

in which they were irradiated to the next cycle phase. The length of the phases of the life cycle is probably, within limits, fairly well fixed and most cell types have a usual or "normal" generation time (the time required for progression from one phase in a life cycle to the same phase in the next life cycle). When cells are irradiated, progression through the life cycle stops and resumes later. This phenomenon is known as "progression delay." Whether there will be progression delay and how long it will be depends on these factors: (1) the phase in which irradiation occurs and (2) the radiation dose.

Usually cells in mitosis (at least those as far as late prophase) complete division, but cells irradiated in the other phases may be delayed. A general observation is that delay is longest the nearer mitosis radiation is given (during mitosis itself, there appears to be no delay). Cells irradiated in early G_1, therefore, would be expected to be delayed a shorter time than those irradiated in G_2, provided the same dose was used. Higher doses cause longer relative delays, but the pattern does not change.

What happens during delay is unclear, but it may well be that cells use the time to repair or recover from radiation damage before progressing through the life cycle. (After all, the end point being evaluated is damage to cellular reproductive systems, and it stands to reason that if the system itself is damaged *while* cells are reproducing, it may be that progress in reproduction is not even possible until the system is repaired.)

The dip or oscillation in split-dose survival curves is thought to result from the phenomenon of progression delay. The first or conditioning dose induces a degree of synchronization in cell cultures. It happens because cells in the most sensitive life cycle phases are eliminated from the *dividing* or reproducing population, either by progression delay or by

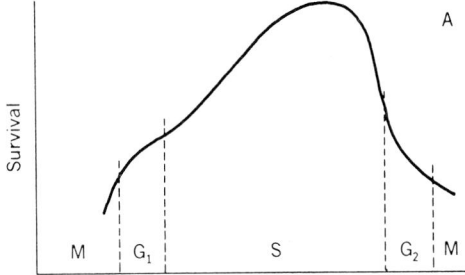

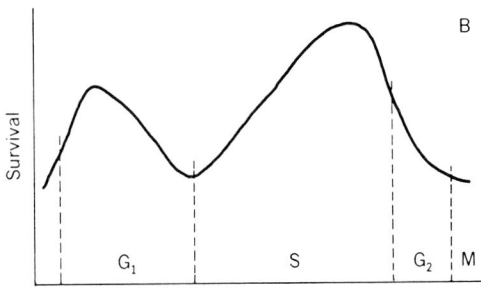

Figure 6.5. Illustrations depicting typical responses of proliferative capacity of mammalian cells following given doses of radiation. A shows the response of cells having short G_1 periods, and B those having long G_1 periods. With some exceptions, these appear typical of most mammalian cell lines tested. For example, in A, cells irradiated in S are most likely to survive; in M, least likely.

death. After irradiation, the first cells to resume reproduction are those that were in the most resistant phases (Figure 6.5).

Those cells which are first fully recovered, then, are those which were in resistant phases when irradiated and are the same cells which are first to begin progression again through their life cycles. Such cells progress from the resistant phases in which they were irradiated and in which they were arrested to more sensitive phases of their life cycles. An effect of irradiation is to restrict temporarily the *reproducing* population to cells in the same phases of life cycle, namely, cells in the most radioresistant phases when they were irradiated. Such reproducing cells progress together from these resistant phases to sensitive ones so that, if reproducing survivors of a dose of radiation are given another dose at an appropriate time after the first dose, the reproducing population will, as a whole, be rather sensitive to this additional radiation. This is so because of the aforementioned synchronization. All or nearly all cells will have progressed to and be in a sensitive life phase. If more time is allowed to elapse, this synchrony is lost and the population begins again reproducing at random. The synchrony in progression from resistant to sensitive phases is the cause of the observed oscillation in recovery or postirradiation changes in radioresponsiveness.

SUMMARY

1. Multitargeted cells may recover from sublethal radiation damage.

2. Recovery requires time, but given enough time, the reproductive mechanism of sublethally damaged cells may fully recover.

3. Recovery appears to be the reestablishment of the target number.

4. Sensitivity to radiation changes as the phases of reproductive cycle change.

General References

Andrews, J. R.: The Radiobiology of Human Cancer Radiotherapy. Philadelphia, W. B. Saunders Co., 1968, pp. 44–49.

Elkind, M. M., and Whitmore, G. F.: The Radiobiology of Cultured Mammalian Cells. New York, Gordon and Breach, 1967.

Hall, E. J.: Radiobiology for the Radiologist. New York, Harper & Row, 1973, pp. 97–131.

Chapter 7

INTRACELLULAR RESPONSES: THE INFLUENCE OF OXYGEN, LET, AND DOSE RATE

7.1 Introduction. The survival of reproductive function of cells is influenced by a number of factors both intrinsic and extrinsic. In previous chapters two intrinsic factors, the capacity of cells to accumulate sublethal degrees of damage and recover from them and the variations in radiation sensitivity of given cells as they move through the phases of their life cycles, have been described. In addition to these, the degree of cellular oxygenation, the LET of radiation to which cells are exposed, and the rate at which radiation is delivered, which are extrinsic or environmental factors, all influence the ability of cells to survive radiation exposure.

7.2 Oxygenation. Oxygen makes cells sensitive to radiation, an effect particularly pronounced when the radiations have low LET. The effect is not confined to single cells, for it has been shown that lesser doses of radiation are required to produce any end point in any kind of organism when that organism is irradiated under conditions of normal oxygenation than when it is hypoxic. This enhancement of radiation effect (or sensitization) is most profound when low LET radiations are used, and as the LET of radiations is increased, it tends to make less and less difference whether normal or less than normal degrees of oxygenation (hypoxia) exist. With high LET radiations, given end points are produced after about the same absorbed dose, irrespective of whether the irradiated subject is normally oxygenated (normoxic) or hypoxic. With low LET radiations, oxygenation can be a very significant factor.

7.3 The Degree of Oxygenation. Words such as *normoxic* and *hypoxic* are relative terms, and efforts have been made to determine the degree of change in radiation sensitivity through a range of tissue or cellular oxygen concentration (Figure 7.1). What is important is the concentration of oxygen in tissues and cells, and in multicellular organisms this depends on its concentration in inspired gas. Air is about 21 percent oxygen and at atmospheric pressure (760 mm of Hg), the pressure oxygen contributes (the partial pressure of oxygen) is 159 mm of Hg (21 percent of 760). This is so because the pressure exerted by a gas in a mixture of

44

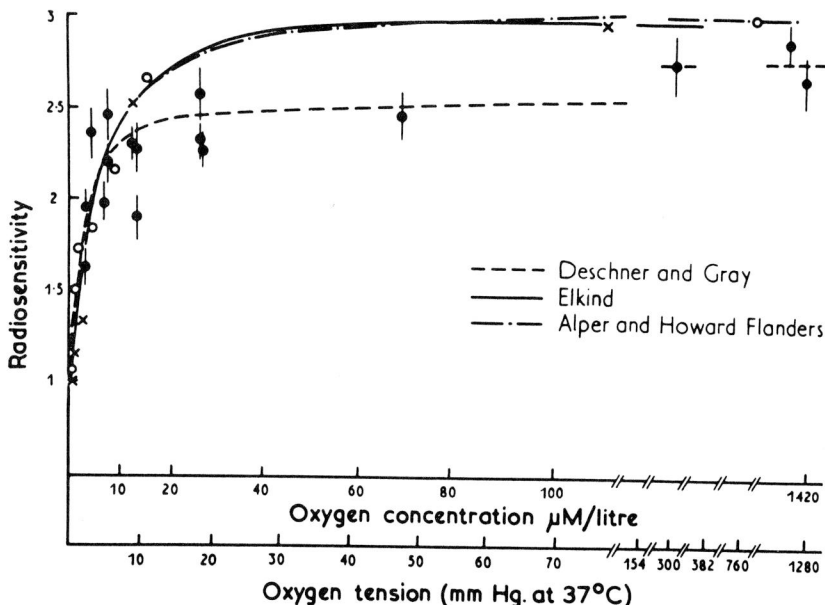

Figure 7.1. Curves of radiosensitivity as a function of oxygen tension at the time of irradiation. Alper and Howard-Flanders, 1956; Deschner and Gray, 1959; Elkind, Swain, Alescio, Sutton and Moses, 1965. (Note: Elkind's curve drawn by Alper from his data. The three curves redrawn and superimposed by Thomlinson.) (From Churchill-Davidson, I.: The oxygen effect in radiotherapy. Oncologia 20, Suppl., 18–29, 1966.)

gases is directly related to the proportion of the mixture it constitutes.

Partial pressure of oxygen determines the concentration of dissolved oxygen in blood, which in turn governs its concentration in cells and tissue. If either the proportion of oxygen in inspired gas or the pressure of the inspired mixture is changed, the concentration of oxygen in blood and tissue is expected to follow suit. Whether this actually happens depends on a number of factors remaining constant in the face of increased oxygen concentration in blood. Among these are circulation rate, the reaction of capillaries to increased oxygen tension, and oxygen uptake by individual cells. It is not certain that these things do remain constant, and some feel the body can compensate for the presence of excess oxygen by stepping down amounts that reach cells. Measurements have been made, and although many indicate increased oxygen tensions in tissues after breathing pure oxygen at either atmospheric or higher pressures,

there are sufficient technical difficulties in making such measurements to cause some investigators to question their accuracy.

7.4 Mechanisms Through Which the Effect Occurs. Several mechanisms to explain the oxygen effect have been proposed, but the following is widely accepted. Molecular oxygen (O_2) has two unpaired electrons. At these sites it readily reacts with many entities that can donate or share electrons with it. Ionizing radiation produces large numbers of free radicals. These have unpaired electrons; to attain stability, they acquire another by reacting with and sharing an electron with something else. The result is that oxygen has a high potential for reacting with free radicals, and the likelihood that key or target molecules will be indirectly affected by irradiation is increased. In this context, "target molecules" do not exclusively refer to the entities represented by extrapolation numbers. Any radiation-produced change in any structure or function must occur as a result of changes in critical molecules. For example, if the end point under investigation is loss of reproductive capacity, the molecules governing that capacity are the targets. If the change under consideration is chromosomal damage, molecules making up chromosomes will be the targets of radiation. Because this is so, a general hypothesis can be developed. Two possibilities exist. Any target molecule may be changed either *directly* by interaction with an ionizing radiation or *indirectly* by the product of an interaction of a radiation and another molecule.

In direct action, the target may be ionized or excited and dissociates. The equation illustrates a general case. RH is a target molecule; R and H are free radicals.

$$RH \xrightarrow[\text{quantum}]{\text{energy}} R\cdot + H\cdot$$

In indirect action, a molecule which is not the target (ordinarily water) is ionized or excited; its free radicals then interact with the target (see Chapter 3).

7.5 Oxygen's Role in Direct Action. Oxygen may interact with the free radicals produced from a target molecule by radiation.

$$RH \xrightarrow[\text{quantum}]{\text{energy}} R\cdot + H\cdot$$

$$R\cdot + O_2 \longrightarrow RO_2\cdot$$

The target molecule, RH, is changed, and RO_2, the resulting free radical, is capable of numerous interactions to satisfy its electronic requirements. Beyond that, a back reaction, which is possible after radiation interaction and would restore the target, may be blocked by oxygen.

$$RH \xrightarrow[\text{quantum}]{\text{energy}} R\cdot + H\cdot$$

$$R\cdot + H\cdot \longrightarrow RH \text{ RESTORATION}$$

$$R\cdot + O_2 \longrightarrow RO_2\cdot + H\cdot \text{ RESTORATION BLOCKED}$$

7.6 Oxygen's Role in Indirect Action. A description of the reduction of oxygen by the hydrogen free radical serves to show one means by which oxygen enhances indirect action.

$$H_2O \xrightarrow[\text{quantum}]{\text{energy}} H_2O^+ + e^-$$

$$e^- + H_2O \longrightarrow OH^- + H\cdot$$

$$O_2 + H\cdot \longrightarrow HO_2\cdot$$

$$RH + HO_2\cdot \longrightarrow R\cdot + H_2O_2$$

RH, the target molecule, is thus changed and raised to a free radical state and becomes capable of a variety of reactions with other cellular substances. This happens despite the fact it absorbed none of the radiation energy itself. *Reactions between free radicals and oxygen increase the probability that target molecules will be changed and lose function as a result of energy lost by radiation in another molecule.*

7.7 Oxygen Enhancement Ratio. The degree of oxygen's enhancement of radiation effects is expressed as the oxygen enhancement ratio, OER. It is determined by measuring the radiation dose required to produce a specific effect under conditions of full oxygenation and the fullest possible hypoxia, and expressing the result as a ratio. There is variation, but the oxygen enhancement ratio of x-radiations and gamma radiations in mammalian cells commonly lies between 2 and 3, normally close to 3. This means that most mammalian cells are about three times as sensitive to x-radiations or gamma radiations under full oxygenation as under extreme hypoxia. But it should also be pointed out that mammalian cells reach nearly full radiosensitivity before they reach their full capacity for oxygen. They increase only slightly in sensitivity with further increases in oxygen concentration (Figure 7.1). For this reason, *slightly* hypoxic cells may not be especially radioresistant. The normal condition for certain kinds of cells (bone marrow is an example) is one of slight hypoxia. But this condition does not confer radioresistance on them.

7.8 Oxygen and Survival Curves. Generally, in the presence of adequate amounts of oxygen, x-ray or gamma ray survival curves have steeper slopes, more pronounced shoulders, and higher extrapolation numbers than the same population under conditions of sufficient hypoxia (Figure 7.2).

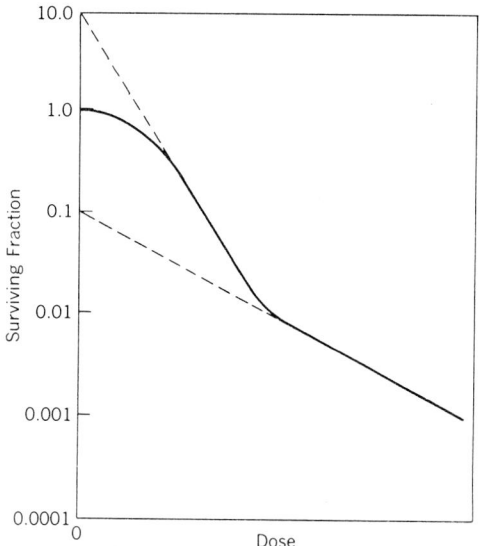

Figure 7.3. A survival curve for a population in which not all cells are equally oxygenated. The fraction of the curve spanning the first two decades represents the response of a well-oxygenated fraction of the population. The later relatively flat portion represents a hypoxic fraction. Extrapolation of the two portions produces two values of n, 10 and 0.1.

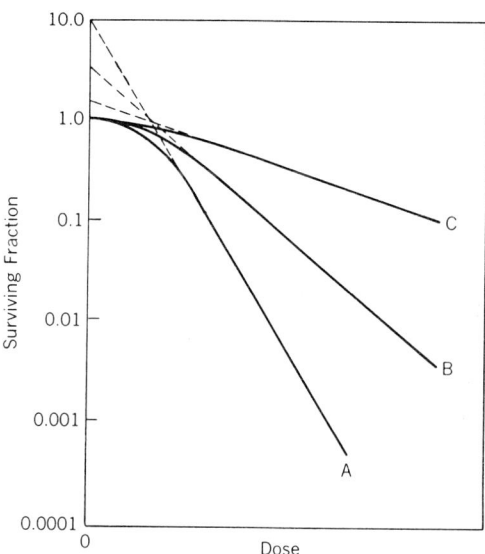

Figure 7.2. Survival of a population under three conditions of oxygenation. Curve A is the survival of fully oxygenated cells. Sensitivity, represented by slope, is greatest and n is highest. Curves B and C represent conditions of increasing hypoxia. Sensitivity (slope) decreases as hypoxia increases, and so does n. Of the three, Curve C is the flattest, has the least pronounced shoulder, and the smallest extrapolation number. It is least sensitive.

The implications are two-fold: (1) Since more radiation is required to produce a given effect under conditions of hypoxia, hypoxic cells must become resistant to radiation. Decreasing slope is a measure of this. (2) The drop in extrapolation number and the disappearance of the shoulder might mean there are fewer targets under hypoxic conditions, but it may also mean less recovery is possible under hypoxia than under normal oxygenation. The latter is usually taken as true. Taken together, these things indicate that while more radiation is needed to inactivate any given target under hypoxia, the mechanism by which damaged targets are repaired or replaced seems impaired, and there is less recovery.

7.9 Populations of Mixed Oxygenation. Survival curves of cell populations in which not all cells are equally oxygenated are inflected rather than smooth (Figure

7.3). The slopes of such curves change at high radiation doses, becoming flatter, indicating a decrease in radiosensitivity. This flatter portion is believed to represent hypoxic cells, and the remainder of the curve, cells of normal oxygenation. Extrapolation of the flat, or "resistant," exponential yields an extrapolation number indicating less than one target per cell in this population. This is unrealistic, and extrapolation of this part of the curve is not believed to reveal target number. Instead, it is thought to represent the fraction of the population that is hypoxic (in this case about 10 percent). Figure 7.4 depicts what is believed to be happening.

In *in vitro* studies, changing oxygen concentration is all that is necessary to change the fraction of resistant cells in a mixed population. There are fewer at high concentrations of oxygen and more at lower. But at the total organism level, this kind of change is not so simple. Extensive efforts have been made to bring about changes in hypoxic populations of cells within total organisms as part of the effort to cure cancers with radiation. Most, if not all, cancers are believed to contain groups of cells that are severely hypoxic. The *number* of these severely hypoxic cells in the cancer is an important factor in the response of cancers to radiation. Because they are resistant, compared to cells in normal tissues and normally oxygenated cancer cells, the more hypoxic cells there are in a cancer, the greater the chance one or more will survive the highest clinical dose with reproductive capacity intact.

Efforts to combat the oxygen effect are critical to successful control of cancer with radiation. These efforts have taken two general forms. The first method increases (1) either the proportion of oxygen in the inspired gas or (2) the pressure of the inspired gas—or both. These have been made with the hope that more oxygen will become available to the cancer *during* irradiation. This supposes that hypoxic tumor regions result principally from poor vascularization which is, in fact, a characteristic of many cancers. Hyperbaric chambers have been built in which patients breathe pure oxygen, commonly at a pressure of 3 or 4 atmospheres, before and during irradiation. But while some clinical advantage has been claimed, spectacular success has not been achieved, and some doubt the technique has much potential.

The second general method has been the use of high LET particles as the source of radiation to minimize the oxygen effect.

7.10 The Influence of LET. Although oxygen is known to enhance all radiation effects, its influence diminishes as LET of radiations increases. The electrons set free by x rays and gamma rays have, on the whole, low LET, and their action is greatly influenced by oxygen. They are about three times more effective in its presence than in its absence. However, as LET increases, OER decreases. The reasons are not known for certain, but the

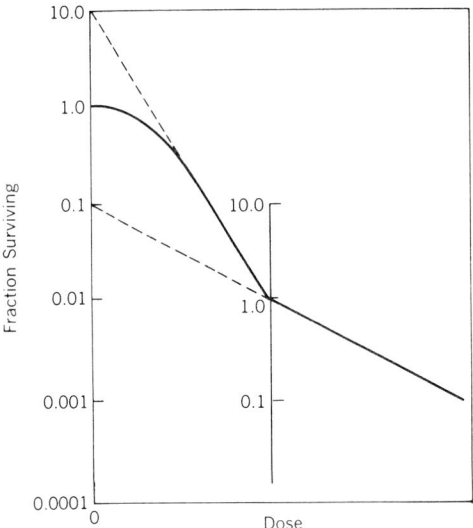

Figure 7.4. A representation showing the response of a cell population of mixed sensitivity to radiation. Of the total, about 10 percent of the population is radioresistant compared to the rest. A set of coordinates erected at the point of change in slope shows an extrapolation number of 1.

effect is believed related to target size and the number of critical interactions that must take place in targets to inactivate them. As stated earlier, targets are molecules or groups of molecules governing specific functions. They are small, and particles of high LET, because they spend much energy in a small volume, have a greater chance of leaving enough in a target molecule to make a critical change. The widely spaced interactions of low LET radiations have a lesser chance of doing that, but chain reactions of radiation-produced free radicals and oxygen tend to increase the radiation-affected volume and the amount of energy ultimately transferred to target molecules. The influence of LET on survival is demonstrated in Figure 7.5.

As LET increases, radiosensitivity increases, and the shoulder tends to disappear, making the resulting curve more nearly exponential. The extrapolation

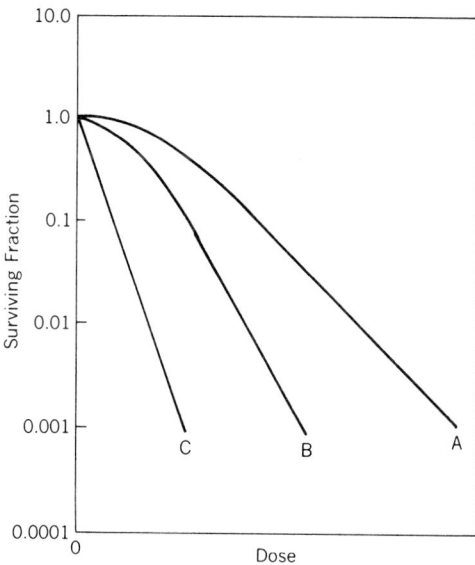

Figure 7.5. The influence of LET on *in vitro* survival curves. All curves represent cells of the same population. Curve A is derived from fully oxygenated cells irradiated with x rays or gamma rays, and Curves B and C are fully oxygenated cells irradiated with particles of higher LET. Of the three, the radiations used to obtain the data for Curve C have the highest LET.

number approaches one. This tendency toward an exponential response is taken to mean that targets hit by high LET radiations are not repaired or replaced. Each interaction is presumed lethal. There is no sublethal accumulation of damage, and consequently, there is no recovery. If these suppositions are true, fractionated high LET radiation doses should give a cumulative response equal to the same total doses given in a single exposure. Studies that have been done to test this point demonstrate that a cumulative effect is possible.

7.11 Dose-rate. The rate at which radiation is delivered can have a profound effect on intracellular response. As a general rule, lesser effects will be produced by any given dose of radiation if the radiation is given slowly, over relatively long periods, than if it is given rapidly. The explanation usually offered is that between interactions with ionizing radiation, biologic systems recover to greater or lesser degrees. The longer the time lapse between interactions, the more complete the recovery, and, as pointed out in Chapter 6, some functions are known to recover completely, given enough time.

Low dose-rate irradiation may be viewed as a kind of dose fractionation. Each interaction with radiation can be thought of as an extremely small fraction with the time lapses between fractions as extremely small recovery periods. As dose-rate decreases, time between these infinitesimal fractions increases, and recovery between each tiny fraction becomes more and more complete. Survival curves reflect this phenomenon (Figure 7.6). As dose-rates decrease, survival curves become more nearly exponential, slopes decrease, and extrapolation numbers approach 1.

The survival curves in Figure 7.6 result from external irradiation, and intracellular response from radioactive materials taken up in cells is not as well known.

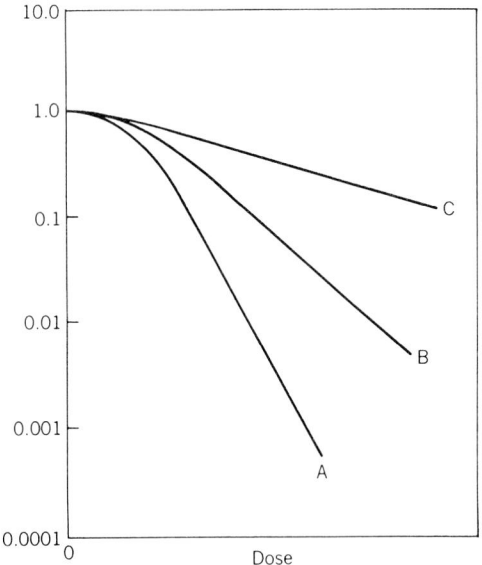

Figure 7.6. The effect on the survival of reproductive capacity from a given range of radiation doses delivered at three dose-rates. Curve A represents the most rapid dose-rate. In B and C the radiation was delivered more slowly, with C having the slowest rate of the three. As dose-rate diminishes, intracellular responses become more nearly exponential and the slope of the curves becomes less steep. The extrapolation number approaches 1.

There is a temptation to expect no differences to exist, but there might be some owing to influences of intracellular site of deposition of various radionuclides, effects of daughters, stable or radioactive, or other factors, but in-

sufficient work has been done to make any generalization possible.

7.12 Dose-rate and Oxygen Enhancement Ratio. Low dose-rate continuous radiation shows a smaller oxygen enhancement ratio than either high dose-rates or large-dose fractionated irradiation.

SUMMARY
1. The presence of oxygen in cells enhances the biologic effects of low LET radiations.
2. The effect occurs only when oxygen in cells is within a definite concentration range.
3. The oxygen enhancement ratio (OER), using low LET radiations, for most mammalian cells is about 3.
4. As LET increases, OER decreases.
5. As LET increases, survival curves show steeper slopes, become more exponential, and have values of n nearly equal to one.
6. With very high LET radiations there appears to be little or no recovery between fractions.
7. Dose-rate influences radiation response. As dose-rate decreases, survival increases. This is believed to be due to recovery between radiation interactions at low dose-rates.
8. OER decreases as dose-rate decreases.

Chapter 8

INTRACELLULAR RESPONSES:
IN VIVO MODELS

8.1 Introduction. Some of the intracellular responses to radiation, namely, the capacity of the reproductive system of certain cell types to accumulate damage at sublethal levels, the sensitivity of cellular targets to radiation, the repair or recovery from sublethal damage, and the influence of cell life cycle on these responses, have all been shown using various cells grown *in vitro*. But most cells do not exist in nature as single entities, living an independent life from other cells and dependent only on the medium in which they are grown for sustenance. Cells are usually part of some tissue, organ, and organism, and thus are part of an extremely complex, interdependent system. It is fair to wonder, then, whether cells in their natural setting, *in vivo*, respond to radiation as do those *in vitro*. That is, does the reproductive mechanism of cells *in vivo* respond the same way as those cells tested *in vitro*? In order to determine this, a number of models have been developed in which the reproductive capacity of cells irradiated *in vivo* could be observed and damage to this system analyzed. Many of the observations made suggest that the response of cells *in vivo* is much the same in character as that of cells tested in tissue culture.

8.2 *In Vivo* Models. Mouse leukemia was one of the first systems to be tested as an *in vivo* model. The model was developed by Hewitt and Wilson[1] and formed the basis of a number of other models developed later. In this model portions of livers of mice which have advanced leukemia and in which the livers have been infiltrated by leukemic cells are removed. These are treated in such a way as to result in a suspension of individual cells in liquid. The number of cells per unit volume in this suspension is determined, and then a known portion of it is diluted with a known amount of appropriate medium. A portion of this second suspension is diluted as was the first, and the process is repeated several times. The result is a series of cell suspensions, serially diluted. From these, inocula are prepared and injected into the peritonea of a number of mice.

The number of leukemic cells required to produce consistently a growth of the tumor in the ascites fluid of a constant fraction of mice is then determined. Usually, the number of leukemic cells required to produce ascites tumors in 50 percent of inoculated hosts (the TD_{50}) is used as a standard end point.

When this TD_{50} has been determined,

the process is repeated, this time using *irradiated* leukemia-infiltrated mouse livers. A series of increasing doses of radiation is given to a number of leukemic mice and the TD_{50} after each radiation dose is determined. Naturally, if irradiation destroys reproductive capacity in some cells, then it follows that fewer cells in each inoculum will be *able* to reproduce and produce an ascites tumor in the host. The TD_{50} will be higher for irradiated cells than for unirradiated cells and should be progressively higher as the dose of radiation is increased. The TD_{50}, then, represents the fraction of cells that have survived irradiation with their reproductive capacity intact. A high TD_{50} is expected after high doses of radiation (many cells cannot reproduce, yet are part of the inoculum); a lower TD_{50} is expected after lower doses of radiation (more of the cells in the inoculum can reproduce and contribute to tumor growth).

Survival curves obtained are like those seen after irradiation of mammalian cells *in vitro* but are characteristic of curves for hypoxic cells. That is, they are exponential and have relatively flat slopes. The reason is that cells grown in ascites fluid tend to be crowded and are indeed likely to be hypoxic. If oxygen is made available to them, however, a curve having a shoulder and a less steep slope, one characteristic of normally oxygenated conditions, is obtained.

Spleen Colonies. Another system for making similar analyses is that developed by McCullough and Till,[2] in which colonies of cells growing in the spleens of mice are used for assay. In this model, mouse leukemia can be used, but so can normal mouse bone marrow.

In the model, bone marrow (either leukemic or otherwise) is removed, usually from the femora of mice, and suspensions, serially diluted and containing known numbers of cells, are prepared. These inocula are injected into mice that have been given massive doses of total-body radiation, in the dose range that is known to kill all of their own bone marrow.

The injected bone marrow cells will populate the bone marrow spaces of the lethally irradiated host mice, and some will grow in other places as well, one of which is the spleen. In the spleen, small nodules appear which are believed to be colonies of cells growing from the injected bone marrow cells. Cells that produce such a colony are known only as colony-forming units (CFU) because it is uncertain which of the many types of cells in bone marrow actually do this. If the experiment is properly carried out, colonies in the spleen are formed relatively far apart and are believed to have been formed from single cells capable of proliferation (single colony-forming units). Each colony, then, represents the progeny of one injected cell which transplants to the spleen and is capable of reproduction. Naturally, when relatively concentrated bone marrow suspensions are injected, there are a large number of spleen colonies. Lesser concentrations produce fewer colonies. The number of colonies depends on the numbers of cells injected. It is a simple matter, then, to determine the *number of colonies* produced by any *given number of cells*.

After the determination of the number of colonies, all that needs to be done is to repeat the procedure using, as bone marrow donors, mice that have been irradiated over the total body. In the case of the donors, however, the radiation doses used are far lower than those given the prospective hosts. The doses are expected to be sufficient to kill *some* bone marrow in the donor and to produce sublethal injuries in many of the others if, in fact, sublethal reparable damage is possible *in vivo*.

A graded series of such doses is given groups of donors, and their bone marrow is transplanted to heavily irradiated hosts. After an appropriate interval, the number of spleen colonies produced is determined. When this determination is done, larger numbers of cells are required to produce the *same* number of colonies produced from injections of unirradiated cells. This result is attributed to the probability that fewer cells having colony-forming potential are able to produce colonies after irradiation and that the number which can produce colonies gets smaller as radiation dose gets higher. In other words, this technique measures the number of bone marrow cells (or colony-forming units) that survive given radiation doses with their reproductive capacity intact.

Results show that, when gamma rays are used, survival does appear to be a sigmoid function of dose; a curve is produced having a shoulder and an exponential component. Bone marrow proves to have a low extrapolation number and a low value for D_0, the radiation sensitivity, confirming that this tissue is indeed radiosensitive.

Crypt Cells. In a model devised by Withers and Elkind,[3] mucosal crypt cells of mouse jejunum are used. A loop of mouse intestine is exteriorized. A portion in the center of the loop is shielded, and the remainder, two regions of loop on either side of this shielded center, is given a high radiation dose. The dose must be high enough to totally ulcerate and destroy the mucosa of these unshielded regions, because they serve later as markers to locate the region between them and also as "sterile" fields, regions in which there are no mucosal cells.

The shield over the center portion of the loop is then removed, and this region is irradiated with one of a number of predetermined doses of radiation. At the completion of this phase of the experiment, a number of mice will have two portions of small intestine that are heavily irradiated and, between these portions, a segment given one of a graded series of radiation doses.

The intestines are returned to the peritoneal cavity and the abdomen is sewn up. After an appropriate time, the animal's abdomen is reopened and the jejunum inspected. The two heavily damaged regions are sought and, when found, are removed along with the less heavily irradiated region between them. The whole is split and opened to expose the inner mucosal surface of the gut.

If properly done, there will be two regions in which the mucosa is totally destroyed and, between them, a region in which the destruction is only partial. The latter region is the one that received one of the predetermined radiation doses. In that region will be a number of nodes rising above the destroyed mucosa. These are believed to have been formed from intestinal mucosal cells that survived radiation with their capacity to reproduce intact.

Animals that have received relatively high doses in this test area will have few such nodes or macrocolonies per square centimeter of mucosa. Those receiving lesser doses will have a larger number of such nodes per square centimeter of mucosa. A dose response curve can then be constructed; the number of nodes or colonies per square centimeter of intestine is plotted as a function dose.

When this is done, using x rays as the radiation source, the survival curves are the same in shape as those obtained *in vitro*. That is, there is a shoulder followed by an exponential region. The shoulder is broad, but the slope of the exponential region indicates that the sensitivity of those mucosal cells is not much different from that of colony-forming units of bone marrow.

Whether the colonies of intestinal mucosa grow from *single* surviving cells in the irradiated field has been questioned. That they grow from cells in the irradiated field and not from mucosal cells immigrating into the region from the healthy, unirradiated gut seems assured by the design of the experiment. This is so because healthy intestine is relatively far removed from the test regions by the bands of destroyed or "sterile" mucosa so that no cells are available to migrate from these sterile areas. Whether the colonies are the result of one or a few surviving in the test region is still not certain, but evidence indicates that they are probably from single cells.

Skin Colonies. Similar results can be obtained using skin colonies. In this technique the hair is removed from a region of the backs of mice. Then, using a superficial x-ray source (the penetration of these x rays is slight) a ring or doughnut of plucked skin is heavily irradiated. All cells in this ring are expected to die and, as in the case of the previously described intestinal mucosal technique, the region may be considered "sterile."

The region within the sterile ring is then irradiated with one of a series of predetermined doses. At the conclusion of this part of the experiment, there will be a number of mice, each having a ring of heavily irradiated, sterile skin and, in its center, a region irradiated with a test dose of radiation.

After a time, nodules, which are descendants of cells that survived radiation with their proliferative capacity intact, grow in the test area. The number of nodules is plotted as the *number of cells* surviving irradiation per given dose *per square centimeter of skin.*

With the technique, the curve is similar to that obtained *in vitro.* There is a shoulder, and the slope of the exponential gives a radiosensitivity comparable to that detected *in vitro.* But in both this and the model using the small intestine, the number of targets (n) can only be inferred, since the *fraction* of *surviving* cells cannot be directly determined. The reason is that no one knows how many cells there are in any square centimeter of either gut or skin and what fraction of this

unknown number can proliferate under any circumstances. Since that is so, it is not possible to know the fraction in which this capacity survives.

8.3 Target Number *In Vivo.* Although it is not always possible to determine directly the target number *in vivo*, it can be inferred by measuring D_q, the designation used for the width of the shoulder of survival curves, and also a measure of the radiation damage that cells can repair (maximum reparable damage). A dose of radiation that produces a given degree of damage (produces a given number of skin nodes, for example) can be given a test system. This dose is called D_1. A second identical test system is given radiation that produces the same degree of damage as in the first system, but the radiation is given in two fractions separated by enough time for repair to occur. The *total* radiation given (both fractions added) is called D_2. Naturally, since repair is presumed to occur between the two fractions comprising D_2, D_2 will be a greater dose than D_1. D_1 is then subtracted from D_2, and the difference will be the damage that was repaired between fractions and should be equivalent to D_q, the width of the shoulder, or the maximum reparable damage. In this way reparable damage can be estimated *in vivo.* When the shoulder width and the slope of the exponential are known, the extrapolation number (n) is easily obtained. The estimates obtained in this way run much higher than determinations of n for mammalian cells *in vitro* and suggest that much more reparable damage can be absorbed when proliferative cells are part of tissues than when they are individuals in an artificial medium. The extrapolation number (n) as determined *in vitro* is ordinarily a rather low number in many of the cell types analyzed, falling between 1.0 and 10.0, with many of the numbers being about 2.0. This suggests that the amount of sublethal or reparable damage that mammalian cells can absorb *in vitro* is relatively small. Determinations made *in vivo*, however, yield values for n often exceeding 100. The reasons for this are unknown. However, the fact it exists suggests that cell killing *in vivo* will be much more difficult than one might be led to believe from data gathered in tissue culture.

Text References

1. Hewitt, H. B., and Wilson, C. W.: A survival curve for mammalian cells irradiated *in vivo*. Nature, *183*:1060–1061, 1959.
2. McCullough, E. A., and Till, J. E.: The sensitivity of cells from normal mouse bone marrow to gamma radiation *in vitro* and *in vivo*. Radiat. Res., *16*:822, 1962.
3. Withers, H. R., and Elkind, M. M.: Radiosensitivity and fractionation response of crypt cells of mouse jejunum. Radiat. Res., *38*:598, 1969.

Chapter 9

INTRACELLULAR RESPONSES:
RADIAION GENETICS

9.1 DNA and the Genes. Genes, the structures which control inheritance and metabolism, are, in nearly all organisms, composed of very large molecules called deoxyribonucleic acid (DNA). These molecules are made up of subunits, joined together end to end, which form two intimately associated chains. Together, the chains make up single, long, un-branched threads, twisted at regular intervals along their length, giving them the form of a spiral staircase (Figure 9.1).

The nature of the linkage of the sub-units in each of the chains and of the chains to each other is extremely impor-tant, for it is the basis of the mechanism by which genetic information and instruc-tions are *coded* and stored in cells, as well as the basis upon which these instructions are transmitted to the cytoplasm. It is also important in the proper carrying of ge-netic information from one cell genera-tion to the next.

9.2 The Composition of DNA and the Arrangement of Its Components. Each subunit of which DNA is composed is itself made up of subunits: phosphoric acid; the five-carbon sugar, deoxyribose; and one of four organic, nitrogenous

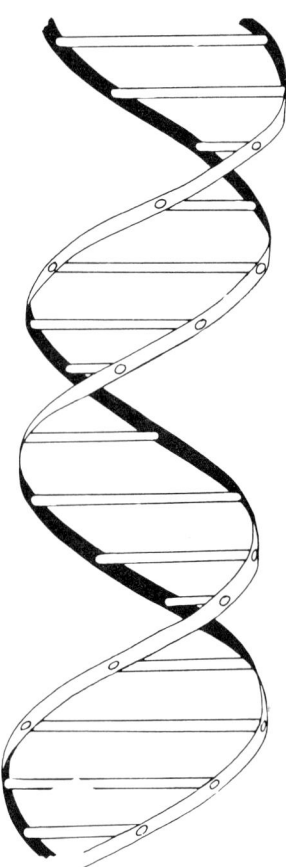

Figure 9.1. An illustration of the configuration of molecules of DNA.

56

bases. Taken together, each of these units is called a *nucleotide,* specifically, a deoxyribonucleotide. DNA molecules consist of two chains of these deoxyribonucleotides, held together by hydrogen bonds, in which the side pieces or backbones of the spiral structure are sugar-phospate and the rungs are nitrogenous bases.

Only four nitrogenous bases are associated with DNA: adenine, guanine, thymine, and cytosine.

The two chains of DNA are alike save for a single, important difference. *Wherever adenine occurs in one chain, thymine will be in the chain opposite; wherever cytosine occurs in one chain, guanine will occur in the chain opposite.* The amounts of adenine and thymine in DNA must be equal; so must the amounts of guanine and cytosine. But there is no necessary relationship between the quantities of adenine-thymine and guanine-cytosine.

9.3   The Genetic Information or Code. It has been stated that cellular metabolism is under control of the genes and that the DNA which makes them up directs all the activities of any given cell. A cell's activities include not only all the chemical reactions going on within it, but will, in addition, include whatever intracellular interactions in which the cell might participate (antigen-antibody interactions as an example).

The functions of DNA are embodied in the sequence of bases along the chains. The sequence constitutes a sort of "sentence," and genetic information or instruction is derived by reading the sentence from the beginning of the chain to its end. Each sequence of three bases, starting at the beginning of the chain, is one bit of information. Taken as a whole, the molecule is a series of bits of information which, presumably, must be in a coherent form for the cell it is in.

The sequences of bases of the DNA chains lead to the production of specific proteins that are the functional and structural units of cells. When the sequence is interrupted and/or thrown off, each group of three *proper* bases is still thought to convey genetic information. But because the sequence is changed, the information is changed. The result is (1) that particular proteins, structural and functional units needed by particular cells, are not synthesized and (2) that something *will* probably be synthesized which does not have the characteristics of the protein that has been lost. Using an analogy, suppose that all the "E's" in the following example sentence were deleted. Suppose also that the reader could only use letter quantities equal to that of the example sentence. Then, FRED GOES TO THE STORE would read FRDG OSTO THST OR. Obviously, the message has been lost and the meaning of the new message is, at the very least, unclear.

Any force, then, that changes the base sequence will produce a *change* in genetic information. Such changes will almost always result in an important change in the metabolism of the cell in which they happen. Specific proteins will not be synthesized, and molecules may be synthesized that are likely to have no bearing on the cell's actual needs.

9.4   The Role of the Gene. The appearance (both functional and physical) of an individual is referred to as his *phenotype* and is a result of the genetic complement, the *genotype.* But, the same phenotype may be produced by different genotypes. Particular functions or physical characteristics are under the dual control of genetic material situated on two separate chromosomes. Such chromosomes are known as homologous; one comes to the genotype from one parent and the other from the other parent. When the genetic material on homologous chromosomes is alike, the function or characteristic it controls is said to be under *homozygous control,* but when it differs it is said to be under *heterozygous control*. In heterozygous situations, frequently the phenotype is an expression of the effect of only one of the homologous genes. The gene that is expressed is called the *dominant* gene, and the gene whose effect is not expressed is called *recessive*. Recessive genes are not ordinarily expressed in heterozygous situations, but dominant genes are expressed, even when carried on only one chromosome.

9.5 Genetic Mutations. Mutation has been known for many years as a naturally occurring phenomenon. Every species of plant or animal studied, including the very simplest, is known to have members that differ from the rest suddenly arise in its midst. These are mutants; they occur after a change in genetic material. Such changes and the mutant individuals they produce appear apparently at random in any population of living things. No *cause* that explains their appearance is now known; as a result, they have been termed "spontaneous mutations."

There is no way to predict when any given gene will change or mutate, but the rate or *frequency* at which spontaneous mutations occur in many genes is known. Many agents increase the mutation frequency. These include chemicals, viruses, and radiations. Of these, ionizing radiation is the most effective agent known for producing mutations in quantity. The production of mutations is, in fact, very likely to be its most important biologic effect. None of its other effects is so far-reaching in consequences *both for the individual in which the mutation occurs and for the population of which a mutant-bearing individual is a member*. For physicians utilizing ionizing radiation as a diagnostic or therapeutic tool, the production of mutations may be the most serious unwanted side effect of radiation.

9.6 The Manner of Production of Mutations: Point Mutations. It has been suggested that in the production of mutations, ionizing radiation acts by ionizing nitrogenous bases in the DNA chains, in particular during DNA synthesis. Pairing in DNA is between thymine and adenine and between guanine and cytosine. But if these molecules are ionized during DNA synthesis, pairing between the wrong bases may occur. As a result, the *sequence* of bases in the newly synthesized chain is wrong, and a permanent, heritable, discrete change in the new DNA results. When the cell which inherits this changed DNA chain prepares to divide, the chain will, of course, direct the synthesis of a new complementary chain which incorporates this mistake. This change is also heritable. Both chains of DNA have undergone a change, and the mutation is complete.

Ionization produced in the bases of DNA *during synthesis* is not the only mechanism by which radiation produces mutations. At *any phase* in the cell's life cycle, base changes or base deletions may occur which alter the critical base sequence of the molecule. Ionization of one or more of the bases, or interaction of one or more bases with free radicals, may so alter their structure that the bases no longer possess the characteristics they had before exposure to radiation. Such a event is a base change, and it will, of course, alter the base sequence (Figure 9.2). The daughter cells that receive these DNA chains will receive, as a consequence, base sequences conveying inadequate or even erroneous information. In addition, this information, whatever it is, will be heritable.

Base deletion occurs if sufficient numbers of ionizations take place within one or more bases or if interaction with free radicals so alters one or more of them that they are deleted from the DNA molecule. The effect upon the cell is, of course, quite similar to that of base change (Figure 9.2). Some genetic information will be missing. Further, what is left does not neces-

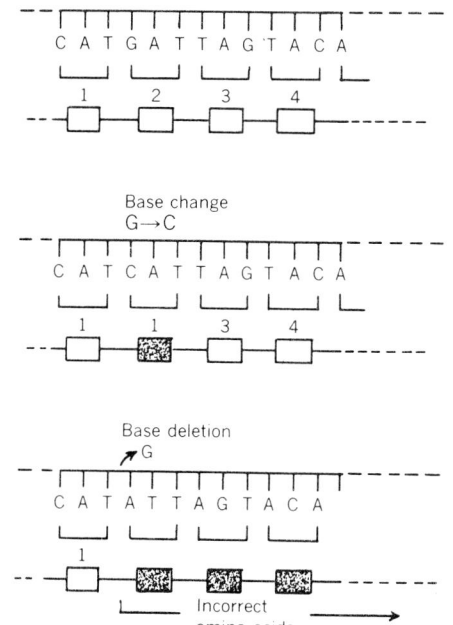

Figure 9.2. The effect of mutagenic agents that change or remove a base. When a base is changed, an amino acid will be changed. Deletion usually results in large-scale changes to nonsense in the code. (Redrawn from Watson.[1])

sarily convey the same meaning to the cell as it did before the deletion took place; what is left can actually have no meaning at all.

9.7 Dependence on LET. Not all the ionizing radiations are equally efficient in bringing about the changes in DNA that lead to mutation. When the frequency of base change and/or base deletion is compared after irradiation with ionizing radiations of increasing LET, it is found that changes in frequency depend upon the LET of the radiation used. The greater the density of ionization produced by radiation (greater LET), the more efficient is the radiation for bringing about mutations.

The dependence of base change or deletion mutations on LET is believed to be due to either the ion density produced within the DNA molecule by the radiation or the quantity of irradiation products

(free radicals) produced in cellular material. As the density of ions per unit volume of matter increases, there is a progressively greater chance that several ionizations will occur within the nitrogenous bases of DNA. There is, therefore, an increasing probability with increasing LET that one or more of the bases will be changed or deleted by ionization or free radical attack. There is a rise in mutation frequency to a maximum with progressively increasing LET, which may represent a rise to a maximum in the number or concentration of mutagenic agents produced by radiation. Such agents could be the primary decomposition products of ionized water (the free radicals H· and OH·) or even the products of interactions among these products which may in turn react with the nitrogenous bases in DNA and change them.

9.8 Mutation Frequency, Dose, Dose-rate, and Stage of the Cell Cycle. The magnitude of increase in mutation frequency is dependent upon dose. The relationship is a linear one in which each increment of dose produces as many mutations or as great an increase in mutation frequency as the one preceding it.

Such a relationship suggests that there is no dose of radiation too small to produce no increase at all in mutation frequency. That is to say, any dose of radiation, regardless how small, probably increases the frequency of mutation; there is *no threshold dose.* At exceedingly low doses, of course, any increase in frequency would be very small, small enough to be at the limits of detection. This fact is responsible for the difficulty in experimentally demonstrating increases in mutations after exposure to very small amounts of radiation and for the near impossibility in ruling out the existence of a threshold dose directly.

There is much evidence showing that the production of mutations by irradiation is independent of the rate at which the radiation is given or of whether a dose of radiation is given all at once—without interruptions—or interrupted in fractions. Individuals experiencing several acute exposures over a period of time show the cumulative effects of these exposures. As a general rule, once established, mutations are permanent except for the very rare occurrence of back mutations. As an illustration of these, male mice that have been bred for nearly 2 years after irradiation show no significant reduction in the number of mutations.[2]

On the other hand, there is evidence of a reduction in the number of transmitted mutations in female mice as the interval between radiation and conception is increased.[3] The frequency of mutation was considerably reduced by allowing an interval of 7 weeks to intervene between radiation and conception. Presumably the lower mutation frequency comes from oocytes in less mature stages at the time of irradiation. The cause of this difference in mutation frequency is not known, but it may involve lower mutational sensitivity of oocytes in early follicular stages, an efficient repair system, cell selection, or a combination of all.

Russell[4] in studies using irradiated mice has also shown that the number of induced recessive mutations varies not only with the *rate* at which the radiation is given but also with the stage of the germ cell. In the highly developed spermatozoa, there was no difference in mutation induction when dose-rate was high (90 R/min) or low (0.009 R/min), whereas studies on the progeny obtained after irradiation of spermatogonia and immature oocytes demonstrated that the number of mutations produced was related to the rate that the radiation was delivered. Thus, although immature germ cells may be capable of repair, there appears to be no repair in mature germ cells, and radiation-induced mutations in them are essentially permanent.

9.9 The Effect of Genetic Mutations.

Mutations are, as a rule, detrimental to the cell or individual that bears them. They bring about sharp deviations from the status quo. The degree of deviation is, of course, dependent upon the number and importance of the genes that are changed during a mutation.

When mutations do occur, the change is most often in a negative direction: that is, from the presence of something to the absence of the same thing. Any gene, presumably, is the bearer of some bit of information, a particular command that should be carried out by the cell in order to have normal function. A gene may, for example, direct the synthesis of a particular enzyme. If the gene mutates, a *negative* change will probably have taken place. In its changed or mutated form it is unlikely that the gene will be able to direct the synthesis of the enzyme—and the gene's "action" will be missing. Part of the deviation from the status quo for this cell, then, will be the permanent loss of an enzyme. In most cases this will be detrimental. Usually the cell will need the missing enzyme; the enzyme's function might be irreplaceable. The only instance under which a mutation is beneficial is the rare one in which there is a fortuitous change in the environment, occurring simultaneously with the mutation, for which the mutation happens to be suited, or if the mutant form is in better harmony with the existing environment. Such events are, however, purely chance occurrences. Mutations are not planned—and they rarely coincide with environmental changes for which they happen to be suited.

All detrimental changes impair *viability*. The result of this impairment is a shortened average life span for individuals that sustain the detriment. In gene mutations, the degree of detriment, which will be reflected as the length of lifetime shortening, depends on the number or importance of the genes changed. Some genes are so important that changes in them bring death essentially immediately. Such changes are called *lethal* mutations. If a lethal mutation occurs in the germ line (the cells giving rise to gametes), the zygote that inherits them will probably be nonviable or, at least, will not survive embryonic life. Other mutations (like the one that gives rise to hemophilia) are *severely* detrimental, yet not immediately lethal. Such mutants may survive into adult life if considerable care is taken to avoid life-threatening predicaments. Still other mutations (like the one that produces albinism) are less detrimental.

Finally, there are mutations that cause such small degrees of detriment that they affect viability very slightly. The life span of these mutants may be almost as long as that of nonmutants.

Given enough time, all detrimental mutations become extinct. There is reproductive pressure against mutants; they are, because of this pressure, "selected" out of a population. Extinction of slightly detrimental mutations may take many, many years; more detrimental ones extinguish themselves sooner. Lethal mutations become extinct almost as soon as they come into existence. The presence of mutations (mutants) in a population imposes a kind of burden upon that population. The mutant-bearing individuals cannot give rise to a line of descendants (except under the chance circumstances where the mutation is beneficial) that can become successful, and so the line is destined for extinction.

Recessive mutations that are highly detrimental or lethal can also have some effect on the heterozygous carrier. On a population basis, the numbers of individuals who are heterozygous for most characteristics may be many times greater than the number of homozygotes; there may be many heterozygotes who are less well adapted to their environment. The effect of recessive mutations may well be felt by many carriers over many generations. One therefore must consider not only the *final* effect of the recessive mutations, the genetic death of a homozygote, but also the effect on heterozygous carriers. The total damage from a mutant gene is the sum of *all* deleterious effects on *all* carriers.

9.10 Ionizing Radiation and Large Populations: Genetic Load. When large populations of any kind of organism are exposed to ionizing radiation, mutations occur in at least some of the cells in some of the organisms exposed. This happens (and has happened from time immemorial) to all organisms, including man. The source of the exposure is the natural background. The amount of radiation to which all matter is exposed (perhaps because it has always been present, perhaps because all things now living have been selected to live in spite of its presence) is not intolerable. The number of mutations that this radiation induces appears compatible with life and the continued success of most species. But exposure to ionizing radiation is increasing. The sources of the added radiation are (for human beings) exposure to medical or dental radiation, exposure as a result of occupation, and (for all living things) exposure to the radiation from radionuclides "falling out" on earth as a result of the testing of nuclear weapons in the atmosphere.

The introduction of too many mutations into a population at the same time can be detrimental, even life-threatening, to the population as a whole. The hazard to man of exposure to industrial and medical radiation is just that; it can threaten his very existence as a species. As industrial and medical uses widen, the risk of mutation for more and more people increases, and the danger to the population increases.

The deleterious genes found within the population make up what has been referred to as the *genetic load*. The size of the genetic load depends on two factors: (1) the rate at which deleterious genes are produced through mutations, and (2) the rate at which they are removed by natural selection. When the rate of removal equals the rate of production, a condition called *genetic equilibrium* is reached, and the level of occurrence of those genes then remains stable over the generations.

Under present conditions of medical care and living, it would seem unlikely that the unfavorable results are being eliminated as rapidly as was formerly the case. In other words, one of the effects of medical progress has been to decrease the severity of natural selection. There is no humane or moral alternative to this.

Even though we cannot alter the selection of these genes out of the population, we can alter the rate at which mutations are produced. Because of the wide use of medical radiation, significant gonadal doses are being given to the population. Levels of increased radiation will add to the deleterious gene pool and will increase the equilibrium level of these genes in the pool. Physicians as a group are naturally preoccupied in treating the ills of their patients. The unborn descendants of these patients, particularly if removed by several generations, may seem remote. Nevertheless, when considering genetic harm from radiation exposure, one *must* attach the same weight to personal injury, no matter how long after the causal event the injury occurs.

9.11 The Doubling Dose. A way that the effectiveness of a mutagenic agent can be expressed is as the *doubling dose*. The doubling dose is the amount of a mutagen needed to produce a number of mutations equal to the spontaneous number. For acute radiation (the type which includes most medical exposure) the doubling dose is estimated to be between 15 and 30 rads.[5] Since chronic radiation has been

found to be less efficient in producing mutations, the doubling dose from chronic, low-level exposure (occupational) is estimated to be about 100 rads. It should be remembered that the doubling dose applies to large populations and is *extrapolated* from animal studies. It is difficult to express genetic risks as they relate to an individual or his family. Further, since most mutations are recessive, their potential damaging effects may not be noticed for generations.

9.12 Chromosomes. One of the principal functions of chromosomes seems to be transport of the genes in proper quantities through cell division. The number of chromosomes in a cell is characteristic of species; all members of a given species carry the same number of chromosomes in all comparable cells. Within individuals, germ cells (the male and female gametes) and somatic cells (the cells of which the body is composed) differ in chromosome number. The gametes have *half* the number of chromosomes found in somatic cells. Germ cells are said to have an "n" number of chromosomes, and the chromosomes which constitute the n number are said to be a *set* of chromosomes. Somatic cells have twice the number of chromosomes as are found in germ cells, but it is more than a simple case of a larger number. They actually have two n sets of chromosomes; that is, each chromosome in the n set of the germ cell is present in somatic cells in duplicate. The symbol employed is 2n. Human male somatic cells have 46 chromosomes arranged in 23 pairs, of which 22 pairs are called autosomes and one pair is formed by the X and Y, or sex chromosomes. In females there are 22 pairs of autosomes plus two X chromosomes. The number 46 is thus the 2n or diploid number for man. In the gonads, the sex cells (spermatozoa or ova) contain the n number, or 23 chromosomes. Each sex cell contains 22 autosomes plus one sex chromosome.

Not all human somatic cells contain the diploid number of chromosomes. There are two reasonably common deviations. (1) Some cells may contain a multiple of the diploid number, a condition called "polyploidy." It is common in the human liver where it is a result of abnormal division. (2) Cells may contain more or less than the diploid number of chromosomes but a number which is *not* a multiple of it. This is known as "aneuploidy."

Sets of chromosomes are vitally important in another way. *Chromosomes are the structures that bear the genes.* They are the bearers of the molecules that direct the metabolism of each cell and, therefore, of the total organism. *In order to have a complete set of genes, in order to have the proper balance between genes, in order for metabolism to be carried out properly, a 2n set of chromosomes is a fundamental requirement.* (Gametes carry out only limited metabolism; they are a temporary state of affairs). Loss of a chromosome or part of a chromosome almost inevitably results in the loss of one or more genes. This can produce metabolic derangement, for, unless another gene is present to take over the function of the lost one, the function itself is likely to disappear. If enough are lost (and even a few genes will constitute enough), life is impossible.

9.13 Chromosome Breakage. The important first step in loss of genetic material is the production of structural alterations in chromosomes by ionizing radiation. This can happen when chromosomes are broken. Chromosome threads are broken either by the passage through them of an ionizing particle or particles or by an attack upon them by the products of irradiated cellular material—usually of irradiated water. Severed or broken ends have the property or capability of joining with and *adhering* to any other fractured or broken end that they happen to meet—usually with the *first* one they chance to encounter. When broken ends meet, the adhesion of fragments causes continuous chromosomal threads to be formed, threads capable of reproducing themselves in exactly the same way as does any intact chromosome.

When a chromosome arm has been broken, it most frequently rejoins the piece from which it was separated. This is true because these segments usually lie close together after a break, and each is, predictably, the first broken end either is likely to encounter. They can and do heal, and recombination of broken ends results in a thread exactly the same as the original unbroken one. This process, reconstituting the original chromosome after a break, is called "restitution."

The restitution of two broken ends of the same chromosome may be dependent upon the stage of division cycle in which a break takes place. The longer a break remains open or unhealed, the greater chance that broken ends of the same chromsome will *not* heal with each other. A broken end can join after a relatively short time with the broken end of another chromosome (provided more than one chromosome in the same nucleus is broken), but it is also possible for fragments to remain uncoupled for long periods of time or even to remain open indefinitely.

The consequences to cells of chromosome breakage, except when restitution occurs, are usually very serious. Exactly what happens and how serious the consequences will be depend upon (1) how many chromosomes in any given cell are broken; (2) how many breaks occur in any given chromosome in that cell; and (3) how many breaks occur within any given arm of a chromosome. These factors are, in turn, dependent upon the dose of radiation given and upon the LET of the radiation that is used. The number of chromosomal aberrations from a single break increases *linearly* with dose, but those resulting from two breaks increase with the *square* of the dose.

A major function of chromosomes is to transmit genes (DNA) through cell division to daughter cells with no loss of genetic matter and in proper sequence. The most serious consequence of chromosome breaks is the chance that fragments of chromosomes will be lost at cell division because broken fragments have no attachment to the spindle. This results in the loss of genes, and daughter cells formed at such divisions will be deficient in this respect. The number of genes lost depends on the size of the fragment lost, which in turn depends on the location of the original break. Nevertheless, the loss of even a few genes can seriously impair normal function and even be lethal.

The loss of the attachment to the spindle does *not* seriously impair the function of the genes in nondividing "resting" or interphase cells. Attachment to the spindle is important only at division and does not bear on gene functions during cellular vegetative life. If cells with broken chromosomes are compelled to divide, then the results of breaks are expressed as unequal division of genetic matter to the daughters.

When more than one break occurs in chromosomes, the results are more complicated. Broken portions of different chromosomes may heal to each other (translocations or insertions) or if there is more than one break in the arm of a particular chromosome, portions may rotate and heal after rotation (inversion). These phenomena result in a changed order of genes which is particularly troublesome during *meiosis*, often resulting in death of cells in the germ line. As a rule such aberrations are not problems in somatic cells, for they do not constitute a serious impediment to mitosis.

9.14 The Relationship of LET to Chromosomal Aberrations. The production of chromosome aberrations by ionizing radiations is dependent upon the type of radiations. Generally speaking, low-LET radiations are less efficient for breaking chromosomes than are high-LET radiations.

In comparisons of radiations of high and low LET for the production of any biologic end point, the presence of oxygen and the amount in which this element is present are of critical importance. This is as true in the production of chromosome aberrations by ionizing radiation as it is for any other end point. The biologic effects of radiations of high LET (alpha particles, for example) are nearly independent of any enhancement by oxygen. But the efficiency of sparsely ionizing radiation for producing chromosome aberrations is altered by changes in cellular oxygen tension. This phenomenon accounts for the great interest in the use of high-LET radiations in radiotherapy. The disorganized chaotic manner of tumor growth results in local variations in the oxygen tension of tumors and in local variations in radiosensitivity. Under these conditions, low-LET radiations have variable effectiveness at killing or sterilizing tumor cells, whereas high-LET radiations are expected to minimize the variability produced by the range of oxygen concentrations.

When all factors are accounted for, it may be properly anticipated that the relative effectiveness of radiations for producing chromosomal aberrations will increase with increasing LET. Maximum effectiveness should be reached, and at very high LET, efficiency may be expected to diminish. This is so because, as particles of increasing LET pass through chromosomes, it becomes more and more likely they will be broken. At the same time, however, as LET values increase, the energy of particles is quickly expended so that they move progressively smaller distances and the probability that they will pass through a chromosome thread then begins to *decrease*. When very high LET values are reached, particles come to rest so quickly that there is only a small chance that they will traverse a chromosome. If one does, however, it will almost certainly break the chromosome.

9.15   Dose, Dose-rate, Dose Fractionation.   The production of chromosome aberrations by ionizing radiation is related to both the dose of radiation given and the time taken to administer the dose. It has been shown that, using x ray as the source of radiation, simple aberrations (one break in one chromosome) increase linearly with increasing dose (that is, each increment of dose produces about the same number of such aberrations as any previous increment) and the number of these aberrations is unaffected by changing the dose-rate or by splitting the total dose into fractions. Such aberrations have been called "one-hit" aberrations to contrast them to aberrations of greater complexity that can occur only after more than one break has been produced in a chromosome. The linear relationship to dose and independence of aberration production from dose-rate or fractionation have led to the conclusion that chromosome breaks occur independently of one another (the existence of one break neither increases nor decreases the probability of another occurring).

The rate at which any given dose of radiation is delivered will be of fundamental importance to the number of *complex* aberrations eventually produced. In order for *any* kind of complex chromosome aberration to occur, it is first necessary for two breaks (whether in one or more chromosomes) to be open at the *same time*. Broken ends do, of course, become restituted, and, if radiation is delivered at a very slow rate, there will be time for some of the breaks to undergo restitution before another break can occur. The net effect on the chromosomes

when radiation is delivered very slowly is that relatively few aberrations of any kind are produced; in particular, there are fewer of the complex ones.

9.16.   Chromosomal Aberrations: The Significance of the Stage in the Cell Cycle.   Both the number of chromosome aberrations and the kind of aberrations produced by exposure to radiation are dependent, in part at least, upon the stage in the cell cycle in which cells are irradiated. In *terms of chromosome damage*, the DNA *postsynthetic stage* (G_2) is more radiosensitive than either the DNA synthesis stage itself (S) or (G_1) the presynthetic stage.[6] Smaller amounts of ionizing radiation bring about more chromosome damage in G_2 than in either S or G_1. Further, the *kind* of aberrations produced is dependent upon the stage in the cell cycle in which radiation is given. If cells are irradiated in G_1, many complex aberrations (dicentric and ring chromosomes) are seen, as well as some of the simple aberrations. If cells are irradiated in either S or G_2, only simple aberrations are detected.

While there is a difference, then, in the "radiosensitivity" of chromosomes which is dependent on the time in the cell cycle at which the chromosomes are irradiated, it must also be said that the chromosomes are sensitive to the induction of breaks and their resultant aberrations at any stage in the cycle. The chromosomes are never immune to radiation damage.

9.17   The Significance of Chromosomal Aberrations.   Since chromosomal aberrations usually involve the loss of significant portions of or even complete chromosomes, the loss of functions that may occur in the affected cell's descendants is usually more far-reaching than those associated with point mutations. Chromosomal aberrations occurring in the cells of the germ line result in genic imbalances that will be immediately lethal for individuals inheriting them or will produce offspring ill-equipped for life. Like those bearing genetic mutations, they are at a disadvantage and are unable to compete with "normal" members of the population. Because aberration-bearing individuals usually deviate more from the genetic status quo than genetic mutants, these individuals will find even minor illnesses and accidents serious hazards, even life-threatening. Even though these individuals will tend to become extinct more quickly than mutants, for the period of time they are members of the population they will be a great burden.

Though the mutations that most affect populations are those occurring in the germinal tissues, such changes also occur in somatic cells. Mutations in these cells are not passed on to progeny, but they may affect the individual. Diagnostic levels of x-irradiation have been demonstrated to produce chromosomal aberrations in the lymphocytes of peripheral blood.[7] As little as 5 R exposure can produce these aberrations, and they can persist for many years. Doses of I-131 (averaging about 5 mCi) are also sufficient to produce such abnormalities. Studies on workers occupationally exposed over years to total doses greater than 10 rads have also shown significant elevations of numbers of aneuploid cells and chromosomal aberrations in peripheral lymphocytes when compared to the population at large.

The significance of these aberrations has not been determined. There is a possibility that there may be a connection between these aberrations and the induction of cancer and leukemia as well as with aging.

SUMMARY

1. All functions of cells are controlled by genes located in the cellular nuclei.
2. The genetic information is codified in the sequence of nitrogenous bases which are part of the DNA molecule.
3. Ionizing radiation increases the frequency of genetic mutations principally by point mutations (base change and base deletion) or by chromosomal aberrations.
4. Exposure of populations to ionizing radiations is detrimental to the population (mainly by the induction of recessive mutations).
5. There appears to be no threshold for genetic damage.

Text References

1. Watson, J. G.: Molecular Biology of the Gene. New York, W. A. Benjamin, 1965, pp. 255–296.
2. Russell, W. L., and Oakberg, E. F.: Radiation genetics of mammals. In Radiation Serology and Medicine. W. D. Claus, ed. Reading, Mass., Addison-Wesley Publishing Co., 1958, p. 189.
3. Russell, W. L.: Recent studies on the genetic effects of radiation in mice. Pediatrics, 41:223, 1968.
4. Russell, W. L.: Evidence from mice concerning the nature of the mutation process. In Genetics Today. (Proc. Internat. Cong. Genet. XI) Oxford, Pergamon Press, 1964, Vol. 2, p. 257.
5. Looney, W. B.: Radiation genetics. In Atomic Medicine. C. F. Behrens, E. R. King, and J. W. J. Carpender, eds. Baltimore, The Williams & Wilkins Co., 1969, pp. 423–424.
6. Hsu, T. C., Dewey, W. C., and Humphrey, R. M.: Radiosensitivity of cells of Chinese hamster in vitro in relation to the cell cycle. Exp. Cell Res., 27:441, 1962.
7. Bloom, A. D.: Cytogenetic effects of low-dose internal and external radiations. In Medical Radionuclides: Radiation Dose and Effects. AEC Conf.–691212, 1970, pp. 425–430.

General References

Spar, I. L.: Genetic effects of radiation. Med. Clin. N. Amer., 53:965,1969.
Watson, J. D.: Molecular Biology of the Gene. New York, W. A. Benjamin, Inc., 1965.

Chapter 10

EFFECTS ON TISSUES AND ORGANS

10.1 Introduction. The responses of organized tissues and organs to irradiation are largely due to the effects of the reproductive mechanism of certain cells in the tissue. The cells of many, though not all, tissues and organs fall into one of three categories or compartments. They are either stem cells, differentiating cells, or mature, functional cells. The stem cells are thought of as immature forms, cells that are undifferentiated and generalized. The purpose of these cells is to reproduce, and, from the stem cell compartment, new cells to replace those lost from the tissue or organ are formed. Of the tissues and organs in which there is a stem compartment, this compartment is usually quite active mitotically compared to the activity of other compartments of the same tissue or organ.

The other compartments are called the differentiating and functional compartments, respectively. In the former, cells originating in the stem cell compartment specialize or differentiate. This means they lose their generalized character and begin to develop the special characteristics they will need in the next state of their lives, when they are functional cells. In certain tissues or organs, the process of differentiation is accompanied by much cell division, but in others, cells differentiate without dividing or only after a few divisions.

Cells of the functional compartment perform the *function* of the tissue or organ for the organism. For example, these are absorptive cells of the mucosa of the gut or some of the circulating cells of the blood. Cells come to this compartment after having differentiated in the differentiating compartment and are regarded as fully specialized and mature. Mature, functional cells rarely reproduce but are replaced when they are lost (they have finite lives) by cells that originated in the stem cell compartment and specialized or matured in the differentiating compartment. Of the three compartments, most mitotic activity is in the stem cell compartment and the least among the mature or functional cells.

Mature, functional cells of tissues and organs are resistant to radiation, for their functions, the functions they perform for the *organism*, seem affected only after very high doses of radiation. This is so probably because the sites performing these functions in such cells are located in the cytoplasm, and the processes involved in the performance of these functions are not easily destroyed by irradiation. Moreover, destruction by radiation of nuclear targets does not affect the actual life span of cells, but does affect their ability to *reproduce* and give rise to a line of descendants. Since functional cells do not reproduce, damage to the reproductive mechanism is not expressed. Therefore, when tissues and organs are irradiated, there is little change in the functional behavior of the mature, differentiated cells and little change in their life span. For a time after irradiation, then, tissue and organ function may remain at or near pre-irradiation levels. The time normal function persists varies among tissues and organs and depends pretty much on the length of life or normal life span of mature cells. So long as tissues or organs have enough mature cells, tissue or organ function remains at or near normal levels. In tissues and organs in which the life span of mature cells is very long or relatively long, therefore, tissue function may remain at or near normal for a very long time, in some cases for the remaining life of the irradiated individual. Examples of such tissues would be nervous and muscle tissues in which the life of mature, functional cells is almost the same as that of the organism.

In tissues or organs in which mature cells have short lives, tissue or organ function may be affected soon after irradition. This is due to direct radiation effects not on those mature cells but on the reproductive mechanism of the immature, reproductive cells in both the stem and differentiating compartments. This system is sensitive, and at any dose some reproducing cells will be made sterile ("killed"), and others will be delayed in progression through their life cycles. The number killed and the length of progression delay will be related to radiation dose, phase of reproductive cycle, and the number of cells in the reproductive cycle when irradiated. Therefore, after irradia-

tion, the supply of new cells emerging from the stem and differentiating cell compartments and bound for the functional cell compartment will fall off to a degree and for a time be determined by radiation dose, number of cells in sensitive life reproductive phases, and number of cells in reproductive cycle. Cells in the mature, functional compartment die and are lost by the tissue at or near the pre-irradiation rate (irradiation does not affect this) but are then replaced at less than the pre-irradiation rate, and a deficit in the number of these cells occurs. With it, a loss in the level of tissue or organ function is also expected, simply because there will be fewer cells to perform it. The degree of deficit and its duration are dependent on those factors already noted above.

It is clear, then, that the tissues or organs that respond most to irradiation are those tissues and organs in which function is most dependent on the reproductive capacity of cells in the stem cell compartment. That in turn depends on the natural life span of mature cells in the tissue. Tissues in which there is a rapid turnover of cells respond early and profoundly after irradiation. In tissues in which the rapidity of turnover is less (the life span of mature cells is longer), the response of tissues and organs tends to occur later and to be less profound.

10.2 The Law of Bergonié and Tribondeau. The Law of Bergonié and Tribondeau was an early attempt to codify the factors going into tissue radiosensitivity. The law states that radiation sensitivity of tissues depends on the number of *undifferentiated* cells in the tissue, the *degree of mitotic activity* in the tissue, and the length of time cells of the tissue stay in *active proliferation*. Active proliferation refers to the number of divisions between the earliest, immature state of a cell and its final mature state. It is clear why the degree of mitotic activity influences the radiation response of a tissue, but it is not as clear why the degree of differentiation is important. For unknown reasons, differentiating cells are radiosensitive and easily killed by irradiation. In well differentiated tissues, there are few differentiating cells and consequently fewer sensitive cells than in tissues containing many undifferentiated and differentiating cells.

10.3 Organ Function. The fate of organ function after irradiation is more complex than that of tissue function. Maintenance of organ function depends on the proper functioning of the various tissues that comprise them. Organs usually consist of a number of tissues. Some, such as vascular tissue, support organ life. Others *contribute* to proper organ function *but do not produce* organ function. Still others perform the organ function proper. For example, the muscle layers in the bowels assure proper movement of food and waste through the bowels and proper digestion and elimination. Yet, digestion of food itself (the organ's function) is carried out by the mucosal cells of the organ, and muscle plays no direct role in this.

If any of the tissues of an organ fail to function, organ function may be impaired or lost, even if the tissue directly performing the function is not damaged or has been repaired. The impairment of organ function then, may occur either early or late after irradiation, depending on the kind of tissues of which it is made. As an example, function of the small bowel is affected soon after irradiation because there is a rapid turnover of the cells of its mucosa, its functional element. The mature mucosal cells have short lives, and replacement from stem cells is necessarily rapid. After low doses of radiation, however, the impairment of function is minimal and of short enough duration that it is rarely fatal. Delay of progression of the mucosal stem cells through their life

cycles ends soon enough so that in time new functional cells will again be produced. The functional cell compartment is repopulated, and function is restored.

On the contrary, function of liver and kidney is only slightly (if at all) disturbed immediately after irradiation unless rather high doses of radiation have been given them. There is a lesser rate of cell turnover in these organs and, consequently, lesser response. Weeks or months after irradiation, however, function may be affected because functional vascular tissue, which supports the organ, is lost at the end of the long natural span of the lives of these cells. These mature cells are not so easily replaced, owing to fatal injuries stored in reproductive cells in the tissue's stem cell compartment. This happens because much lethal damage (damaged chromosomes are examples) is expressed only at or after cell division. Lethally damaged cells in the reproductive compartment may not die but remain as part of the compartment, being eliminated only at division. In tissues like vascular tissue, the demand for new cells is relatively low and damaged cells in the stem cell compartment remain a long time. When mature cells are needed, the supply is then curtailed because mitosis in the stem cell compartment produces fewer viable cells than normal. The function of this tissue is reduced and so is that of the organ it supports, but the reduction occurs long after irradiation.

SUMMARY

1. Loss of tissue or organ function after irradiation depends upon the length of life of mature, functional cells in the tissue or organ.
2. If the life span of functional cells is long, the function of tissues and organs will persist for long periods uninterrupted.
3. If the life span of functional cells is short, the function of tissues and

organs may be impaired soon after irradiation because they cannot be replaced fast enough owing to radiation damage to reproductive and differentiating cells of the tissue or organ.

4. The function of any tissue or organ may be impaired later after irradiation because damage to tissues supporting the life or functions of other tissues or an organ may be expressed then.

Chapter 11

EFFECTS IN THE TOTAL ORGANISM: THE IMMEDIATELY LETHAL EFFECTS

11.1 Introduction. The principal effect of exposure of the whole body to penetrating ionizing radiation is the shortening of the life of the exposed organism. The length of time life is shortened is dependent upon the dose-level to which the organism is exposed. (It will also depend upon various other factors such as species differences, age of the organism at the time of irradiation, sex of the irradiated organism, time in a circadian cycle at which radiation is given.) It is possible that there are *very* low doses of total-body radiation that do not elicit the earlier death of the irradiated animal, but this is not proved. Most evidence tends to indicate that if such a threshold does in fact exist, it will be very low indeed.

11.2 Effects of Dose. The observed response to single exposures of ionizing radiation given uniformly over the total body is primarily dependent on dose. For any similar group of mammals, exposure to increasing doses results in the appearance of a growing number of signs (those commonly observed are nausea, sometimes with vomiting; hair loss; loss of appetite; malaise, a general but undefined feeling of being unwell; soreness in the throat; petechiae, tiny, local hemorrhages into tissue; diarrhea; and weight loss, in some to the point of moderate emaciation) indicating that response to the radiation is occurring. But ultimately, a dose-level is attained at which, in addition to the above signs, some of the animals begin to die. The dose-levels at which any particular sign appears or at which the first deaths occur will not be the same for all animals. There will be significant species variations (Table 11.1), and, even within species, significant variation may occur.

Table 11.1. The Approximate Lethal Dose for 50 Percent of the Irradiated Animals in 30 Days ($LD_{50/30}$) for Various Organisms

Organism	$LD_{50/30}$ in Rads
Pig	250
Goat	250
Dog	250
Burro	300
Man	300–450 (?)
Monkey	500
Mouse	600
Rat	700
Rabbit	800
Frog	700
Newt	3000

Nevertheless, for nearly all mammals, and for most other vertebrates, the observed constellation and pattern of signs with increasing dose will be similar.

As dose is increased beyond that which begins to kill some of the irradiated animals, more animals will die, and survival time of the animals that die grows shorter. But deaths at any dose-level (high or low) will occur randomly distributed about a mode (there can, at certain dose-levels, be two modes). That is, at any dose-level, death of any particular irradiated animal may occur at any time (long or short) after irradiation. In practice, however, it has been observed that the greatest number of deaths will occur, for any given dose, at a particular time (a certain number of days or hours) after irradiation. The time when the greatest number of deaths occurs will be the mode. It will, of course, vary with species and with other variation-producing factors, but for groups of similar (if not identical) animals, the modes for given dosages will occur at nearly the same time after irradiation.

When mean or average survival times (means are distinct from modes) are computed for each dose of radiation greater than that which begins to cause death, a pattern, consistent for nearly all species of mammals studied, emerges (Figure 11.1).

The pattern of response has three clearly distinguishable components. Initially, over a dose-range of 200 to about 1000 rads, the response is dose-dependent; mean survival time, as dose is increased, decreases from weeks to days. The second phase, extending over the wide range of approximately 1000 to 10,000 rads, is a plateau. The response is independent of dose; mean survival at any dose in that range is about 3 to 4 days. Finally, the last component of the pattern

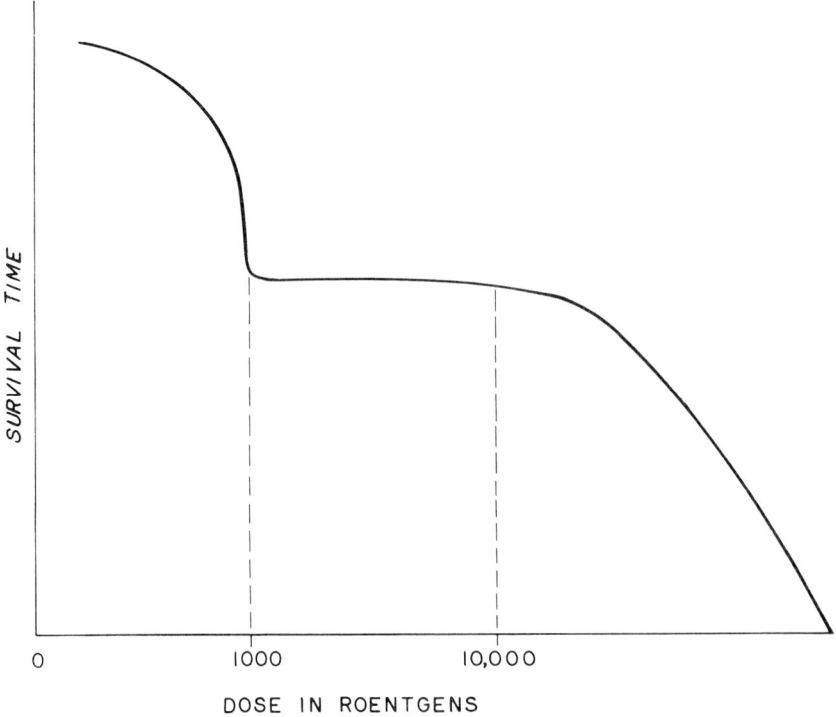

Figure 11.1.   Diagram depicting the *mean* survival time for mammals following a single dose of radiation to the total body.

is again dose-dependent. As dose increases, mean survival time decreases from a period of days to hours and, finally, even to minutes. In the dose-ranges where response is dependent on dose, variation in the length of time that individual organisms survive can be quite wide. In the dose-independent range, on the other hand, there is usually little variation. Because of the variance in survival of individual animals at any given dose of radiation, it is not possible to predict with absolute accuracy how much time will elapse after radiation before all the animals in an irradiated group die (if indeed, all do die). Because some animals in an irradiated group may live for a long time, they are not representative of the group as a whole. As a result, the immediately lethal response from a single dose of total-body radiation is often described as the dose required to kill a certain fraction of a given group of irradiated animals within a given period of time. Thus, a dose of radiation may be expressed as the $LD_{50/30}$ (the lethal dose for 50 percent of the animals within 30 days of radiation), or $LD_{50/18}$ (the lethal dose for 50 percent of the animals within 18 days of radiation), or, of course, any similar expression. This method of evaluation is used in order to minimize the influence (in dose-response studies) of animals that may live for extended periods after irradiation. It will also be a reflection of variation-producing factors.

11.3 The Radiation Syndromes. The three regions of the dose-response pattern (Figure 11.1) are widely believed to reflect radiation damage to, and the failure of, three different organ systems after radiation. In the first region, death may occur within weeks to days; it is believed to be due to the result of radiation damage to and failure of the hematopoietic system (the organ system responsible for manufacture of the corpuscular elements of the blood). This is not to say that other organ systems are unaffected by their exposure to radiation; many cells in many organs and organ systems will be damaged by total-body exposure. Also, in that range of dose-levels, many cells in organ systems other than the hematopoietic system will have sustained enough damage so that the organ or organ system will also fail to function (for example, the gonads and the lymph nodes are badly damaged in that range of doses), but death is due to the failure of the hematopoietic system—the organ system whose proper function is vital to the organism as a whole. Animals can survive without gonads and lymph nodes. The syndrome (the complex of signs and, in human beings, symptoms) preceding death is the result of radiation damage to many, if not all, of the organs in the body, but the hematopoietic system is the radiosensitive system whose failure brings about death.

Therefore, at the lowest immediately lethal doses (those at which some, although not all, animals die, usually within 30 to 45 days after total-body radiation) death may be ascribed principally to the failure of the hematopoietic system. As the dose is increased, more animals die, and the mean survival time grows shorter and shorter. However, as dose is increased, all organ systems (and in fact, all cells) are more greatly affected. A dose range is finally reached in which large numbers of cells of the gastrointestinal tract are badly damaged by the radiation. When enough are damaged, death of the irradiated animals will occur principally as the result of damage to and the failure of that system to function properly. This occurs in the region of dose independence. Mean survival time for most mammals will be about 3 to 4 days following radiation (for man, the period is estimated at approximately 6 days), and there will be little variation about this mean.

Finally, in the dose-range over 10,000

rads, mean survival again varies with dose. In this region, death is due principally to the damage to the blood vessels supporting the central nervous system resulting in its failure. This syndrome is referred to as the cerebrovascular syndrome. Although death is due to the failure of the central nervous system, all other organ systems will also be seriously damaged. The gastrointestinal and hematopoietic systems will both be severely damaged, and they will fail, just as they do at the lower doses. But the failure of the central nervous system brings death very quickly (in less than 3 days) so that the consequences of failure of the other systems do not have time to express themselves.

The shift from one mode of lethality to the next is not a sharp, well-circumscribed thing. Rather, as dose is increased, more and more cells of the next most radiosensitive vital system (hematopoietic–gastrointestinal–cerebrovascular complex) will be damaged. Totally irradiated animals in the dose-region near the shift from one syndrome to the next may die from the failure of either of the systems, depending on the animal itself. Some animals will die from the failure of one of the systems; others will die from the failure of the other (Figure 11.2).

There will be two modes of death, then, following irradiation, one occurring sooner after irradiation than the other. If other signs are present to confirm it, this is usually interpreted as the animals in each mode dying as a result of the failure of different organ systems. The responsible system is not too difficult to identify in the transitional region between bone marrow and gastrointestinal syndromes. The signs accompanying each of them are rather different, and the survival times from each of the two are relatively distinct. In the transitional area of dose between gastrointestinal death and cerebrovascular death, on the other hand, it is often impossible to state with certainty whether the failure of one or the other organ systems has been responsible for death of the animals. All of the signs of the gastrointestinal syndrome are usually present as, of course, will be those characterizing central nervous system failure. The time of death (or the survival time) is often the only practical parameter that can be used to separate the two. Animals surviving 2 days or less are more or less arbitrarily said to have died from cerebrovascular failure while those that live longer than 2 days are said to have died from failure of the gastrointestinal tract.

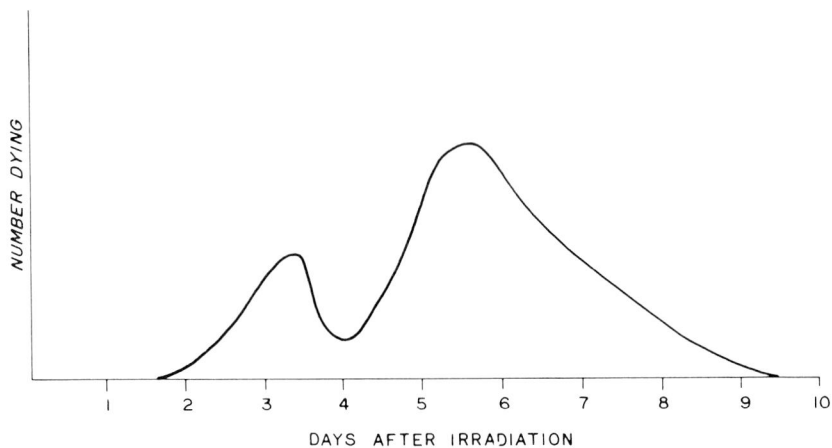

Figure 11.2.   The pattern of survival (or death) following a single dose of total-body radiation at a dose-level near the transition between two syndromes. There are two modes.

11.4 Age. There is a general relationship between radiation sensitivity and age of the irradiated organism. Generally speaking, radiosensitivity decreases as age increases; older organisms are more resistant than young ones. But, the relationship is a general one. Very old organisms become relatively more radiosensitive than younger ones. This increased sensitivity, however, may be a reflection of the loss of resistance to any form of insult (not merely radiation) on the part of the very old. Throughout adult life, with the exception of the period of old age, radiosensitivity changes little. During the period of rapid growth to maturity, most organisms will be more radiosensitive than those having adult status. Puberty is an especially sensitive period.

11.5 Sex. In some species, and in some strains within species, a sex difference in response to total-body radiation has been noted. In mammals, females appear generally to be somewhat more resistant to radiation than males, but the differences are not great.

11.6 Body Weight. Although the precise role of body weight is unclear, it appears that heavier organisms are more resistant to radiation than are lighter ones of the same kind.

11.7 Temperature and Metabolic Rate. The development of radiation injury is dependent on metabolic rate. Because this is so, varying the metabolic rate will bring about variations in the rate of development of radiation injury. If metabolism is speeded up or caused to race (this can be done by subjecting animals to exhaustive exercise or placing nonacclimatized animals in cold environments), the lethal effects of irradiation are enhanced. Retarding the metabolic rate (this can be done by lowering body temperature, inducing hypothermia) protracts the period over which radiation damage occurs. The administration of thyroid-stimulating substances or the removal of the thyroid gland, procedures that will change the basal metabolic rate (although in opposite directions), does not, surprisingly, have much effect on radiosensitivity.

11.8 Diurnal Variation. The sensitivity of some organisms to radiation varies according to the time at which they are irradiated. The effect, described so far only for total-body radiation in rats and mice, is similar to many responses of organisms to the effects of a large number of stimuli; they occur with a daily rhythm. Maximum sensitivity occurs during the active period (the "subjective day").

11.9 Stress. Radiation is more effective (or more damaging) to animals experiencing some kind of stress. For example, small nonlethal doses of radiation may cause death in animals suffering a nonlethal thermal burn. Radiaton appears to be a stress itself and, combined with other stresses, attains greater magnitude of effect.

SUMMARY

1. Irradiation over the total body significantly shortens the life of the irradiated organism.
2. The degree of shortening depends on dose, but at high enough doses death occurs "immediately"—from a period of weeks to hours after radiation is given.
3. The immediately lethal action of total-body radiation, dependent on dose, can be ascribed to the failure of a vital organ.
4. The order of sensitivity of the vital organ systems to radiation, progressing from the most sensitive to the most resistant, is the hematopoietic system, the gastrointestinal system, and the cerebrovascular system.

Chapter 12

THE HEMATOPOIETIC SYNDROME

12.1 Introduction. The hematopoietic or bone-marrow syndrome occurs following a *single* exposure of radiation (requiring a few minutes to, at most, a few hours to give) distributed *uniformly* over the whole body. Death occurs in at least some, though not necessarily all, of the irradiated animals within a few weeks after irradiation. The source of radiation must be an external source of either x or gamma rays (or other *penetrating* radiation, such as might be obtained from a particle accelerator), for, while the bone marrow itself might be affected, even *selectively* affected by an internal source of radiation, a *uniform* exposure of the total body is not likely to result (Appendix B).

12.2 Manifestations. In all species studied, disturbance in the function of the gastrointestinal tract occurs within a few hours after irradiation. This disturbance, which may persist for a few hours to a few days, consists of feelings of nausea, sometimes accompanied by vomiting, and a modification in gastric emptying time. The period of gastrointestinal distress is called the *prodromal* period. It is followed by a period of variable length called the latent period, in which the irradiated organisms are free of signs (in human beings, free of symptoms as well). It must be stressed that this latent period is not a period of inactivity. Actually during this time, the most important consequences of radiation exposure leading to its lethal effects are in progress (the beginnings of the destruction of the bone marrow). This period is "latent" only in the sense that the irradiated organism apparently feels well and is, on the surface, relatively sign- and symptom-free.

The latent period is itself followed by a period of severe illness. In it, signs of gastrointestinal disturbance reappear. Diarrhea, often bloody, sometimes quite severe, begins. Hemorrhage into the tissues occurs, fluid imbalance accompanies it, and, ultimately, serious infection. These are the signs, then, that precede death and lead to it. At death it can be seen that nearly every organ in the body has been affected. Death comes, of course, as a result of damage to all the organs, but principally as a result of the failure of one, the *bone marrow.*

Concomitant with the above signs, several processes will be in progress.

These are also the result of irradiation, but they do not lead directly to death. Among them will be weight loss (the loss of weight is actually a good sign that radiation has been received), the loss of hair (this, too, is a common response to irradiation), and depression of spermatogenesis or oogenesis. These, however, are not part of the syndrome that leads to death. They occur at the same time but are unrelated to the syndrome.

12.3 Histologic Changes. Most species show the *same* histologic changes in the bone marrow following total-body irradiation in the low dose-range. The degree (or severity) of the response (the number of cells responding and the intensity to which they respond) depends on dose. And, of course, the dose that elicits the response in the first place will vary with species.

Almost *immediately* after irradiation (depending on dose—the first few hours to days) the structure and architecture of the bone marrow are disrupted. The marrow normally consists of *nucleated* cells, a small amount of fatty material, and some circulating blood in vessel-like channels (it is the blood supply of the marrow itself). After irradiation it becomes *rapidly* and *noticeably* reduced in the amount of cellular material. The decrease in the number of cells is compensated by dilation of the blood sinusoids (the vessel-like channels), by *hemorrhage* of blood (extravascular red cells) into the cell-depleted regions, and by an increase in the amount of the fatty substance. Ultimately, if the dose is high enough, cell depletion becomes severe; the bone marrow can become entirely acellular, and the marrow space filled with pooled blood. Circulation of blood does not entirely cease (hemorrhage has not so disrupted the channels that they carry no blood at all), but it does become very sluggish and reabsorption of extravascular blood goes on only to a limited extent.

Of course, every cell in the marrow will not be killed; it is doubtful that even at extremely high doses all are killed. Some start to undergo mitosis and an attempt at regeneration (repopulation of the bone-marrow space) will begin. At higher dose-levels (and, as always, the actual dose-level depends on species) these first attempts at regeneration may fail or *abort*. The cluster of dividing cells regresses and disappears. Later in time regeneration will again begin. Precursors to red and white cells will appear after a general *hyperplasia* that seems to characterize the beginning of this more "real" regeneration.

The start of regeneration, even of true regeneration, does not indicate that an irradiated animal will escape the immediately lethal effects of its experience. Regeneration of bone marrow may begin, even after fairly high doses of radiation, even in animals that will soon die from that exposure to radiation.

Lymphoid tissue (in mammals, the lymph nodes and the thymus gland) is severely affected in the dose-range that produces the hematopoietic (bone-marrow) syndrome. Shortly after a dose of total-body radiation, the nodes become severely depleted of cells, and node architecture is completely disrupted. Regeneration of the nodes occurs, usually soon after radiation and, depending on dose, can be rapid. As opposed to effects in the bone marrow, cellular depletion of lymph nodes and thymus is much less dependent on species (less dependent on LD_{50}). A given dose of radiation affects lymphoid tissues in many species of mammals to about the same degree.

12.4 Cytologic Changes. The *number* of cells in mitosis in the bone marrow falls precipitously after irradiation. This is followed by a rise, which may occur at varying times after irradiation (the steepness or rapidity of that rise depends on dose), that results in an "overshoot" of the normal frequency. There is another fall, returning to a level below that of the normal frequency, followed, in turn, by a second rise.

Although the mitotic frequency is eventually restored to the normal unirradiated frequency, it does not mean that the cells that divide will be normal. To the contrary, many will have been injured, and they can be very abnormal. Large numbers of abnormal-looking mitoses are not, however, often observed in *regenerating* bone marrow of irradiated mammals. It is, of course, possible that the damage to chromosomes is quickly repaired in these cells, but the more likely event is that the injured cells are rather quickly eliminated (see Chapter 9 on the fate of cells in which mitotic or chromosome abnormalities have been produced by radiation). Some few cells with damage to the chromosomes, do, predictably, survive (the genes affected were not important enough to have caused their immediate death) and continue to multiply. These, in animals that survive the hematopoietic syndrome, can be detected many years later.

Within a *few hours* after irradiation and well *before* bone-marrow regeneration (of course, the exact time depends on dose and on species), a range of *nuclear* abnormalities of cells in the bone marrow can be seen. The nuclei of some of these cells (for the most part those cells which will give rise to red cells) will be shrunken, their chromatin will be clumped and will stain heavily (pyknosis). The nuclei of others (also red blood cell precursors) will appear very faint, as if their chromatin were dilute or dissolving. These cells are dying (if they are not already dead); the injury from radiation has killed them even before they were able to divide. Still others will have abnormal or aberrant chromosomes in bizarre-looking nuclei. Nuclei may be swollen; there will be chromosome bridges, broken chromosomes, and cells in division in which the spindles have more than two poles (obviously, daughter

cells of such divisions—there will be three or more—will have unequal numbers of chromosomes), as well as other grossly abnormal-appearing cells.

Most of the directly damaged cells (the pyknotic ones or those in which the nuclei are undergoing dissolution) are quickly eliminated and will be in the marrow for only a short time after irradiation. But those with slightly injured chromosomes may persist for some time in irradiated organisms. They, too, will eventually be eliminated.

The *elimination* of cells directly following radiation is probably responsible for the loss of total cellularity in marrow of irradiated organisms. Of course, not every cell is destroyed, and, in some, no obvious damage can be seen. But this does not mean that no damage has occurred. It is reasonable to expect some of these survivors to have sustained mutations. It is also to be expected that some change in the time and even the length of the mitotic cycle will have occurred (mitosis will almost certainly have been inhibited in some).

12.5. The Latent Period. These degenerative changes in the bone marrow, as well as the attempts at early (though abortive) regeneration, go on during the prodromal and the latent periods. The latent period has been described as a period of relative well-being, but, as can be seen, the *important* changes, those degenerative processes in the bone marrow which will lead to death of the organism, are in progress during this period. For death, as a result of the hematopoietic syndrome, is believed to be due to the loss or failure of the bone marrow to carry out its function—that is, supplying the organism with the functional cells that it needs in its circulating blood. The direct loss of some red and white cell precursors due to injury from radiation, the later loss of some of the remaining from injury to their nuclei (mutation and chromosome aberrations), and the inhibition for varying periods of time of the mitoses of some of those cells that escape immediate death from radiation will leave, in a short time, a critically short supply of circulating cells (the degree of that shortness of supply depends on dose). This is true for several reasons. For one, circulating cells have limited life spans. Most are highly specialized, carefully adapted for a particular function, or, at most, for a few functions, which they carry out until they are worn out or become useless. They do not divide and perpetuate themselves but depend instead upon cells in the marrow to supply the organism with their successors. The cells of the bone marrow, then, are regarded as "stem cells," cells that are themselves undifferentiated but that divide frequently, giving rise to cells that are differentiated, specialized, and functioning. These stem cells are the cells that radiation directly eliminates. The cells circulating in the blood wear out or end their lives as do those same cells in unirradiated animals, but none, or very few (the actual number depends on dose), are produced in the marrow to replace them and preserve the function they perform for the organism. The loss of that function by the organism (if such a loss persists too long) can result in the organism's death.

12.6 The Blood Counts. The loss of precursor cells in the bone marrow is reflected by the number and kind of cells in the circulating blood. Changes from the usual numbers of cells found in circulating blood and from the numbers of individual cell types usually present will indicate when the supply of these cells is running short. After radiation, there is a drop in blood count; its severity depends on dose. If the animal will be a survivor, regeneration of the blood-forming elements of the marrow will go on, so that

survivors from the immediately lethal effects of radiation will show an eventual increase in the numbers of circulating cells.

The *anemia* (loss in the level of circulating blood cells) following total-body radiation in the hematopoietic syndrome dose-range is *not* due entirely to the lack of production of red cells by the marrow. A drop in production of red cells does occur, of course, but this is not initially very great and the length of life of circulating red cells is rather long. Red cells are lost from circulation, however, by several other mechanisms related to radiation exposure. Breakdown of the wall of capillaries in many species causes large numbers of small hemorrhages into the surrounding tissue (petechiae). Some, though not all, of these cells return to circulation so that a loss of red cells (the degree of severity of this kind of loss appears highly species-dependent) from the circulation can occur this way.

Another, more important mechanism is a deficiency in the number of *platelets* (a blood cell type involved in clotting) which results in a significant increase in *clotting time* after irradiation. No change in the actual clotting mechanisms or of the efficiency of the clot itself has been shown, but the time required before a clot *is* formed is longer in irradiated than in nonirradiated animals. This time factor is believed due to the loss of platelets and, indeed, the injection of *fresh*, intact platelets into irradiated animals corrects the defect.

12.7 Infection. In both human beings and lower animals, infections are associated with the terminal phases of the radiation syndrome. In human beings (evidence was gathered from the survivors of the atomic explosions at Hiroshima and Nagasaki), fever is common, and at death evidence of general infection is found. This is also true of large animals; they experience a temperature rise (even to levels above 105° F), and overwhelming infection can be seen in many organs. In small animals (mice) bacteremia is present (in this condition bacteria are present in the blood stream). It is distinguishable from the situation in man and large animals in which infection, though very severe, is not as widespread. However, the methods used to detect bacteremia have differed between mouse and man. That used in mice is more sensitive. It is possible, then, that bacteremia may occur in man and large animals, but that it goes undetected.

The bacteria isolated from infected animals are all from the normal intestinal flora of the animals. This is as true of large animals and man as it is of mice.

Infection by these bacteria *greatly* influences the time course of survival of irradiated mammals. Pretreatment with antibiotics (so as to "sterilize" the intestine) will bring about a longer mean survival time if pretreated animals are later irradiated. Moreover, if lethally irradiated animals are treated with antibiotics soon after irradiation, their mean survival time is longer than that of animals not so treated. Finally, germ-free animals survive radiation doses that would kill like animals that are not in the germ-free state.

It appears, then, that there is an increased susceptibility to infection in totally irradiated animals compared to that in those not irradiated. This susceptibility increases *both* with respect to organisms *not* normally pathogenic and those that normally are pathogenic.

Infection, then, has an important part to play in mortality. The length of survival, as well as the dose that may be immediately lethal, is determined by infection. The precise cause of infection after total-body radiation is not known, but it is *correlated* (though not *proved* to be causally related) with the loss of circulating blood cells, namely, granulocytes. With the initial severe drop in these cell types directly after irradiation, there is some mortality. Mortality decreases during the *abortive* regenerative period.

SUMMARY

1. Total-body radiation, if given in sufficiently low single exposures, brings death within a few weeks.

2. While many organs and tissues are damaged and death is due to the damage in all of them, death will come about *principally* as a result of damage to the bone marrow.

3. The time course of survival and the signs and symptoms that accompany it are called the hematopoietic syndrome. In its initial phase the obvious manifestation is nausea, sometimes accompanied by vomiting, but it is at this time that the lethal effects are beginning and undifferentiated stem cells in bone marrow start to die. New cells for the circulating blood are no longer produced. The next major manifestation is a period of apparent well-being, the latent period. But, during this period, more precursor cells in the bone marrow die, and the marrow spaces become nearly cell free. As time goes on, because circulating cells are not renewed, a drop in blood count is observed. At the same time a period of severe gastrointestinal disturbance begins (diarrhea, later becoming bloody); it is the result of radiation damage to the gastrointestinal tract., Hemorrhage into tissue, fluid imbalance, serious infection, and, ultimately, death follow.

4. Death is the result of the failure of the bone marrow and the body systems that combat infection. However, if infection is artificially controlled, these organisms will still die as a result of failure of the bone marrow.

General References

Bond, V. P., Fliedner, T. M., and Archambeau, J. O.: Mammalian Radiation Lethality. New York, Academic Press, 1965.

McLean, A. S.: Early adverse effects of radiation. Br. Med. Bull., 29:69, 1973.

Patt, H. M., and Brues, A. M.: In Radiation Biology, Vol. I, part 2, Chapter 15, A. Hollaender (ed.), New York, McGraw-Hill, 1954.

Chapter 13

THE GASTROINTESTINAL SYNDROME

13.1 Introduction. When mammals are exposed to single doses of penetrating ionizing radiation in amounts greater than 1000 rads, mean survival time of the irradiated animals reaches a constant number of hours or days over a wide range of doses (up to approximately 10,000 rads). Survival time is about 3.5 days for most mammals tested (there are, of course, species variations), but, regardless of precisely how long the survival time is, the important fact is that it will be *independent* of dose. Since survival time due to failure of the bone marrow is clearly dose-*dependent*, it is reasonable to assume that, when survival time is *not* dose-dependent, death is not due to the failure of the bone marrow alone and that another system or other systems are primarily involved.

13.2 Manifestations. The signs and symptoms that characterize the syndrome are the result of radiation damage to and failure of *two* organ systems: (1) the cells lining the intestinal tract (the *mucosa*) and (2) the cell renewal system of the bone marrow. The true or complete *gastrointestinal* syndrome can be obtained only after *total*-body radiation; irradiation of the gastrointestinal tract alone can be fatal, but the full syndrome must involve the cell renewal system of the bone marrow as well.

After irradiation, in all species of mammals tested, there occurs a lack of appetite, sluggishness and inertia that increase with time, diarrhea, and signs of infection and dehydration. Animals irradiated will lose weight, voluntarily take in little food or water, retain the food and water in the stomach for long periods, and, with time, will absorb less and less of the food eaten. At the same time, the numbers of circulating white blood cells will drop to nearly zero.

The last or terminal phase of the syndrome is over rapidly (often lasts for less than a day). It is characterized by diarrhea, which becomes very severe with time, vomiting, and complete exhaustion. The increasingly severe diarrhea brings severe dehydration, noticeable weight loss, and distinct emaciation. Irradiated animals become inert, lying or sitting at rest, moving scarcely at all. Finally at death, which will occur in most species with regularity between 3 and 4 days after irradiation, blood volumes will have changed, electrolyte levels in the serum will be altered, and there may be evidence of bacteremia (bacteria in the blood). There are, of course, species variations in the mean length of survival following irradiation; man and monkey present interesting variants, for mean survival in these species is closer to 6 days than to three or four.

Autopsy of animals of any species shows severe wasting and dehydration. The stomachs of some show retention of food and water. The small bowel itself may be swollen and may contain a bile-colored, liquid material, generally bad-smelling and often tainted with blood. The large bowel may contain liquid stools that often are bloody. There will be large quantities of mucus in the gut as well. The mucosal layer of the small intestine will be badly damaged: villi will be flattened and shrunken, there will be areas that lack the cellular lining altogether, and many of the crypts, which are usually active in cell proliferation, will be empty (Figure 13.1).

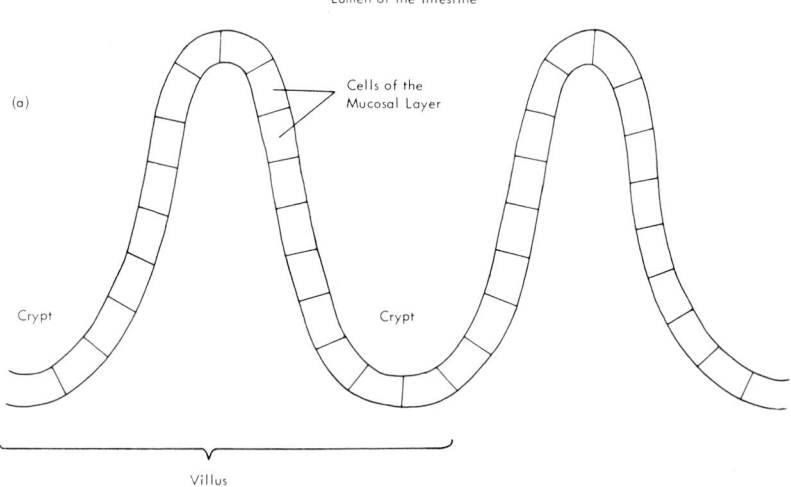

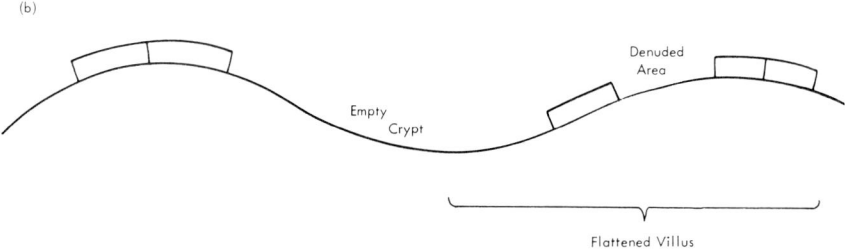

Figure 13.1. The appearance of a segment of small intestine before and after irradiation in the amounts leading to the gastrointestinal syndrome. The normal shape (columnar epithelium) of the absorptive cells in the mucosal layer is distorted; these cells become thin and stretched.

In the stomach and rectum, some areas will be denuded of cells and will be ulcerated. The bone marrow is formless, completely without structure.

13.3 Degenerative Changes in the Lining of the Small Intestine. At the dose-levels producing the gastrointestinal syndrome, changes, severe in nature, appear soon after irradiation in the epithelial lining (the mucosa) of the small intestine. Changes also appear in these cells following much lower doses of total body radiation (they can be seen after doses of 100 rads), but few cells are involved, and the changes produced in the gut itself seem to be minimal. As dose is increased, more cells are, of course, involved, but intact or unaffected cells in the gut divide rapidly to replace them so that *regeneration* more than adequately compensates for cell loss. At much higher doses the degeneration of the intestinal mucosa is clearly seen and can be described.

13.4 Histologic Changes. Soon after irradiation, usually during the first day, progressive destruction can be seen in the *nuclei* of the cells that line the crypts. The nuclei of these cells become pyknotic (the chromatin agglomerates, its amorphic structure indicates that it has condensed into a formless mass, and it becomes densely staining), both nuclei and cytoplasm swell, and, in some, there is dissolution or liquefaction of the damaged cells. These cells are dead (or dying) so that a measurable decrease in the number of cells in the crypt lining follows. Those cells that remain have abnormal-appearing nuclei and are often swollen or enlarged. Few cells in mitosis are seen (and the crypts of the small intestine are usually *very* active mitotic centers), and those that are seen will be abnormal (the aberrancies of mitosis discussed in earlier chapters are present). The debris of the dead cells, which becomes detached from the crypts, begins to build up in the crypt

lumina. In time, however, this is passed down the intestine and disappears. With time, more and more crypt cells die and are sloughed. Few mitoses occur so that these cells are not replaced, and soon regions of the crypts appear that are substantially stripped of cells. Those few cells that remain usually appear grossly abnormal.

The villi themselves lose cells, progressively more and more of them with time, and begin to shorten or shrink. The rate of cell loss and shrinkage is dependent on dose; it occurs faster at higher doses of radiation. At death, these villi are nearly flat and almost completely free of cells.

The precise time schedule of these events and the time required before the intestine is entirely denuded of cells varies with species. In mice, rats, dogs, and goats this condition is reached between 3 and 4 days after irradiation, but in the monkey or in man it probably does not occur until the fifth postirradiation day. The *sequence* of events, however, is the same for most mammals tested (there is some question whether the hamster or the guinea pig, which have a somewhat different pattern, die from a gastrointestinal syndrome of the same sort as that which affects other mammals).

13.5 Regeneration. As in the case of bone marrow, not every cell in every crypt of the small intestine will be killed by irradiation, even at high dose levels. Of those that remain, some will be very damaged and unable to reproduce, but others will have escaped enough damage and are able to reproduce themselves. This is not to say, of course, that they have been unaffected by their exposure to radiation. It merely means that the capability of reproduction has not been destroyed in them. They will form the nucleus or a focus of regeneration of the gut. Simply stated, an attempt to *regenerate* a new gut lining *can* be made. At doses of radiation well into the range that produces a gastrointestinal syndrome, death usually comes to the irradiated animal before any significant regeneration is possible. But, at lower dose levels near the transition range between the bone marrow and gastrointestinal syndromes, uninterrupted regeneration does occur. There is an increase in the mitotic rate, the epithelial lining of the crypts begins to be repopulated from cells within them capable of division, and these cells begin to move into the villi to replace those which have been sloughed. New nests of cells can appear which are thought to represent new crypts.

At still lower doses (those in the range producing the bone-marrow syndrome), regeneration of the intestinal lining is more rapid, and an intact lining may again be produced. Unfortunately, of course, repair of the intestinal lining is no insurance against eventual death brought about by loss of function of the bone marrow.

13.6 The Large Intestine. The remainder of the gastrointestinal tract undergoes a pattern of changes similar to those observed in the small bowel. The main difference is that the changes occur more slowly so that, when death comes follow-

ing a gastrointestinal syndrome, the large intestine is rarely completely denuded of its epithelium. This is to be expected; it is in accordance with the renewal requirements and turnover of mature cells in the large and small bowels.

13.7 The Effects on Cell Renewal. The mature, differentiated, functional cells of the gut epithelium are those lining its villi, in particular those at the tips of the villi. The crypts, much like the bone marrow, are a generative center. The cells in them are undifferentiated and unspecialized. The specialized functional cells of the villi (in particular, those of the small intestine) have a short life. In the normal course of events and under normal circumstances they wear out and are sloughed into the lumen of the intestine only a few days after they have become functional. The gut lining, then, undergoes rapid renewal. The cells in the crypts undergo cell division often and are the source of replacement of the rapidly renewing epithelial lining. There is an orderly procession of cells from the crypts up the villi through a region in which they mature, through another in which they function, to the villus tip where they are extruded and sloughed (Figure 13.2).

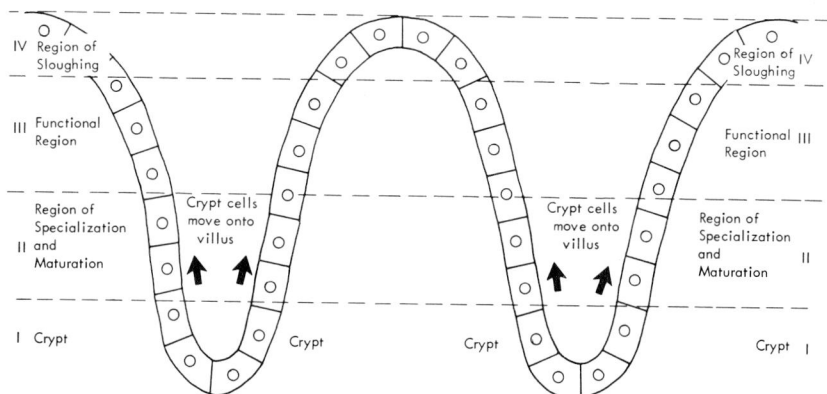

Figure 13.2. Schematic diagram of the regions of a normal intestinal villus. In region I, the crypt, there will be a high mitotic rate in the unspecialized, undifferentiated cells that compose it. Cells produced in the crypts move up the villus to region II where they begin their maturation and specialization. They move up into region III after having specialized in region II and begin to function. They continue to move up to the villus through the functional region, and at the tip (region IV) are extruded or sloughed.

As in the case of the bone-marrow syndrome, radiation strikes at the cells that are the source of the differentiated functional cells (those responsible for the proper function of the organ). Much as in the hematopoietic syndrome, this is to be expected, because it will be these cells that will be most frequently active in mitosis and synthesizing DNA, so that many of them will be vulnerable to radiation damage.

13.8   Effects on Mitosis. Cell division, usually high in the small intestine, may be expected to drop after irradiation as those cells going into mitosis are inhibited. In the species tested this was indeed the case. Over a wide range of doses (75 to 900 rads in the rat; 35 to 325 rads in the mouse), cells in mitosis all but vanished in the small intestine within 30 minutes after exposure. The length of time mitosis was inhibited depended on dose. After inhibition had passed, a rise in the number of mitoses occurred, followed by a second drop or depression in the mitotic count. The initial rise in mitoses, as well as the length of the second depression in mitosis, is also dose-dependent.

Many of the cells that do survive to divide have nuclear abnormalities that kill them. There will be chromosome lagging, stickiness, and broken chromosomes. These cells, often immediately, sometimes after one or two divisions, die.

13.9   Precursor Cells. The crypts of the epithelial linings of the bowel are generally regarded as the generative region for the mature cells of that lining. The relationship, however, unlike that of the bone marrow to circulating cells of the blood, has been difficult to establish. Morphologically these precursor cells, unlike precursors of circulating cells, are not different from cells in the region of maturation (the cells in region I, Figure 13.2, cannot be *morphologically* distinguished from those in region II). Nor, as

can be done for bone marrow, is there any way at present to determine whether the generative cells (precursor cells, "stem cells") have in fact lost the ability to proliferate (the bone marrow of irradiated animals may be transplanted to other animals where the proliferative capacity can be evaluated in the absence of the complicating factors of the bone-marrow syndrome). It is not possible, then, to obtain *directly* an estimate or measurement of the sensitivity of these cells to radiation and to compare them—for that characteristic—to bone marrow cells. But, an indirect mechanism has been devised. It is based on the *assumption* that death *is* due to a reduction in precursor or stem cells below a critical number. Death, then, results if radiation kills or inactivates enough stem cells so that their number is either exactly equal to or less than this critical number. Thus, animals are irradiated using two dose schedules: (1) a single lethal dose or (2) two doses (split-dose) sufficient to kill the organism. Dose-survival curves are then plotted, and the information obtained permits some observations about the epithelial precursor cells. The dose required to reduce the number of such cells to any particular fraction of the normal is much higher than the dose required to reduce the number of bone-marrow cells by the *same* fraction. One would predict (on the basis of the above assumption) that bowel cells are less sensitive than bone-marrow cells and that, if lethality following from the failure of these two organ systems results from the reduction of the number of stem cells to the same critical value, then the dose required to kill 50 percent of irradiated mammals in the gastrointestinal syndrome range will be expected to far exceed that for bone marrow. Experimental observations are in agreement with this prediction.

13.10   The Bone Marrow. Cells in the bone marrow, at the dose-levels required to produce a gastrointestinal syndrome, will, of course, be badly damaged by their irradiation. And damage to bone marrow is an integral part of the *full* gastrointestinal syndrome. The bone-marrow renewal systems are completely depleted within a few days after irradiation at the gastrointestinal syndrome level. In particular, the granulocyte count in nearly all species tested reaches *very* low levels within 3 days of irradiation following doses over 1000 rads. There will be a severe lack of those cells during the very period in which the signs and symptoms of the gastrointestinal syndrome are reaching their apex. A depletion of other types of cells originating in the bone marrow does not occur (in spite of the fact that the bone marrow itself is completely depleted) simply because death comes long before these circulating cells have had time to become exhausted. In these high dose-ranges, there is little change in red cell count. The bleeding that does occur is related to ulcerated and denuded sections of gut. Unlike the bone-marrow syndrome, frank hemorrhage does not have time to occur.

13.11   Fluid and Electrolytes. Animals in the last throes of the gastrointestinal syndrome become very obviously dehydrated. Near death, dehydration results in a profound reduction of body fluids so severe that the blood becomes exceedingly thick, so thick in fact that it is difficult even to withdraw a sample. Electrolyte loss and electrolyte imbalance had been presumed to play a significant role in the gastrointestinal syndrome. Recent evidence casts doubt on the role of electrolyte loss in rodents. After doses of 2000 rads, rats lost only 10 percent of the Na^+ body pool in feces and intestinal contents over 4 days. Therapeutic application of Na^+ solutions did not benefit the animals. The fluid loss *can* be accounted for by the reduction of intake of food and water resulting from the loss of

appetite associated with this syndrome. However, the severe diarrhea, loss in the efficiency of intestinal absorption, and leakage of fluid into the lumen through the damaged intestine are the factors principally responsible for fluid and electrolyte loss.

Although fluid and electrolyte loss is not great in the early phases of the gastrointestinal syndrome, with the onset of severe diarrhea, serious fluid loss begins. During this terminal period, the fluid stool increases both in quantity and in frequency, in time even becoming hemorrhagic. This diarrhea, the cause of the fluid loss, is probably brought about by the failure (or the inability) of the distal end of the small intestine to resorb bile salts and the resulting irritation of the colon by these salts. Bile excreted into the small intestine normally is resorbed by the distal ileum. But, this segment of gut, stripped of functional absorbing cells due to sloughing and the failure of the renewal system in its crypts, does not resorb the bile. The salts go through into the colon, irritating and causing stool to be defecated.

If the bile duct is tied off in irradiated animals, or if the bile flow is diverted so that it does not go into the intestine, this will prevent the diarrhea. But liquid bile or bile salts introduced into the intestine of irradiated animals in which the bile duct has been tied off or diverted will produce diarrhea. Equal volumes of ordinary saline solution will *not* give the same effect.

13.12   Infections. Animals irradiated to dose-levels resulting in the gastrointestinal syndrome may be expected to become infection prone. Indeed, infection does appear to play a major role in causing death at these dose-levels. Irradiated animals are in a poor nutritional state; lymphoid tissue, including that of the intestine, is destroyed (in many species within 3 days of irradiation). The intestine is denuded of mucosa. This allows bacteria in the intestine to gain access to underlying tissue. Antibiotics are of some value in extending the life of the lethally irradiated animals. Germ-free animals survive doses and live longer than do conventional animals irradiated to the same dose. Fever has been shown in dog and in man irradiated to these levels. All these things point to the role played by infection, and it appears to be a significant role.

SUMMARY

1. The full gastrointestinal syndrome is brought on only by *total-body* exposure.
2. Death is the result of damage to many tissues, but the most important of these is the gastrointestinal epithelium and the renewal systems of the bone marrow.
3. Death itself is due to fluid and electrolyte loss, infection, and nutritional impairment. The irradiated animal dies in shock.

General References

Bond, V. P., Fliedner, T. M., and Archambeau, J. O.: Mammalian Radiation Lethality, New York, Academic Press, 1965.
Gitus, J., and Gerber, G. B.: Electrolyte loss, the main cause of death from the gastrointestinal syndrome? Radiat. Res., 55:18, 1973.
McLean, A. S.: Early adverse effects of radiation. Br. Med. Bull., 29:69, 1973.
Sigdestad, C. P., et al.: Intestinal crypt survival: The effects of Cobalt-60, 250 KVcp X-rays, and fission neutrons. Radiat. Res., 52:168, 1972.

Chapter 14

THE CEREBROVASCULAR SYNDROME

14.1 Introduction. Mean survival time following radiation doses that bring on a gastrointestinal system syndrome is rather constant and, with increasing dose, is dose-independent. But, dose cannot be increased indefinitely without bringing about a change in the length of the mean survival time. At some dose-level (the threshold for this effect is not particularly well defined) and at doses above this level, mean survival time will again become dose-dependent. The "threshold" for this effect is, in most species tested, in excess of 5000 rads and, from this dose (whatever it might be for any given species) to doses so high that death occurs even while the radiation is being given, the signs and symptoms elicited are characteristic of damage to the central nervous system. Accordingly, the syndrome of effects has been labeled the cerebrovascular syndrome.

14.2 Manifestations. The course the syndrome follows includes periods of agitation alternating with remarkably apathetic behavior; these manifestations give way to disorientation (in species where this can be evaluated), upsets of equilibrium, loss of coordination of muscular movement (ataxia), diarrhea, vomiting, tetanic spasms of the muscles of the back (the head and lower limbs bent backward and the trunk arched forward), convulsive seizures, prostration, coma, and death.

In some species (among them, monkey) one or more of these signs may occur during irradiation. The time of the onset of the others and their progression depend to a certain extent on dose. During the irradiation itself, animals can become exceedingly active and irritable. Also during irradiation, these manifestations can be followed by apathy; in some cases the degree of apathy can be profound. Occasionally, there also will be vomiting (in animals that can do so), salivation, diarrhea, and nystagmus (oscillatory movements of the eyeballs).

The foregoing can, in a sense, be regarded as a prodromal period, for after irradiation there is a period (variable in length) in which the irradiated animals may appear to be normal. Even during that period, however, there is a sign characteristic of irradiated animals: minimal *voluntary* movement or activity is observed.

Following this phase (which can, dependent to an extent on dose, last only a few moments) other signs make their appearance—uncoordinated movement, random, undirected movements, and tremor. Irradiated animals may go into convulsions even in response to stimuli that usually do not provoke such reactions. Tremors, convulsive seizures, rolling of the eyes, vomiting, repeated evacuation of the bowel, diarrhea (usually quite watery), and a hysterical state, resembling meningitis, are often observed, sometimes alone, sometimes in combination. Prostration and coma follow; breathing is labored; there is gasping, and blood pressure reaches the low levels observed in shock. This is the terminal period of this syndrome; it is the most prominent stage in the syndrome. The most characteristic or remarkable attributes are the tremors and convulsions, circling when the animal is attempting to walk, and, even when the animal is lying down, walking movements in the extremities.

14.3   Histologic and Inflammatory Changes. Blood cells, in particular granulocytes, mononuclear types, and macrophages, filter into the meninges (the membranous coverings of the brain) following the very large doses of radiation that produce the CNS syndrome. Vasculitis (inflammation of blood or lymph vessels) is a common finding only a few hours after a high radiation dose. Veins and arteries of all sizes undergo changes which involve infiltration of granulocytes in foci about the vessel. In the cerebrum this is a striking alteration; the cerebellum, brain stem, and spinal cord seem less involved. While these alterations appear first in gray matter, ultimately white matter becomes most seriously involved.

The inflammation of the meninges, beginning as granulocytic infiltration, commences very early, increases to a maximum, and falls off minimally during the remainder of the course of the syndrome.

The choroid plexuses are quickly affected by the radiation. Liquid infiltration (edema) and infiltration of leukocytes occur soon after irradiation. They reach a peak and then diminish, but they are still detectable late in the course of the syndrome. The reason for these reactions is unknown, but it has been supposed that they occur in response to the production by radiation of minute, nonseptic areas of tissue damage. Such areas do exist (microhemorrhages and necrosis have been found in the nearby tissue), and have been taken as evidence supporting such a view. The infiltration occurs so quickly that bacteria, as causative agents, may be ruled out. Radiation damage to the tissue itself and/or damage to capillaries resulting in leakage from them and foci of pressure which injures and kills tissue are the probable causative agents.

14.4   Cytologic Changes. Pyknosis of the *granule neuronal cells* of the cerebellum is a characteristic finding following total-body (and head) irradiation. The nuclei of these cells take on a more intense stain and shrink in volume. These changes, while dose-dependent, occur very soon after irradiation (at higher doses more cells will be affected and more irradiated animals will show this kind of damage), and increase with time. They do reach a maximum, however, and in some, if not all, species appear reversible.

The mechanism underlying shrinkage and pyknosis in these cells has been represented as the shifting or migration of fluid out of the nucleus. There *are* changes in specific gravity and water content of the brain after irradiation which may tend to substantiate this view. But if this is the case, the reason underlying such fluid movement can be ascribed *either* to a direct action of radiation upon the cell nucleus or to a response of the cell membranes (a change in permeability), or that of the capillaries (again, a permeability change), to irradiation. The observations made in the brain support either view.

In the remainder of the central nervous system, few morphologic changes have been observed following *total-body* exposure. Those that have been seen can as well be attributed to postmortem changes or to changes resulting from edema as to radiation itself. It is possible that all changes at the histologic level may not become morphologically apparent during the short survival time.

14.5 Consequences of Vascular Damage.

The damage to blood vessels (both large and small) already cited (vasculitis) and changes in capillary permeability permit leakage into the brain itself of the substances usually confined to the blood vessels in the brain. There is, however, disagreement as to whether the vascular changes observed are the primary consequences of radiation in this dose-range that lead to death. While some feel that these are the critical events, others believe them to be quite incidental. Nevertheless, by permitting extravasation of fluids and dissolved components into brain tissue itself, the astrocytes (the cells some believe to be responsible for maintaining the blood-brain barrier) can be damaged, and the *blood-brain barrier* may have broken down. Aside from this, however, edema which results from vessel damage *is* responsible for some local cellular damage.

In some species, after irradiation of the head only and following irradiation of the total body, there is evidence of some gross edema of the brain. There is an *increased* water content of the brain in those whose heads only have been irradiated. Such an increase has not yet been shown in totally irradiated animals. If such changes in fluid balance do indeed occur after total-body irradiation, they may have resulted from damage to the vascular system.

14.6 The Immediate Causes of Death.

While the cause or causes of death are not known, it is known that the cerebrovascular syndrome is truly a *total-body* syndrome. Death is *presumed* to result from events taking place within the skull (the brain itself, the meninges, blood vessels; everything contained *within* the skull), but irradiation of the head alone does not produce death when it is given in the same dose-range as that which elicits death from the cerebrovascular syndrome; much higher doses directed to the head alone are required to induce death.

Lack of evidence of large necrotic areas in the brain suggests that, while the neurons *may* have been lethally irradiated, they do not appear to have died or to have stopped functioning at the time death comes to the total organism. Death, then, is not likely to have come about from a loss of the neurons' function.

The increase of fluid content of the brain may cause the organ to swell, even if only slightly. The bony confines of the skull will not permit much expansion without resulting in a buildup of pressure in the brain. Swelling in irradiated tissues is not unusual; it has been noted to occur transiently after irradiation in both tumors and normal tissues.

All these observations suggest the possibility that damage to the cells of the brain due to an increase in intracranial pressure, damage to the blood vessels of that organ, and edema are the mechanisms of death resulting from the cerebrovascular syndrome. But this has not as yet been proved.

SUMMARY

1. Mean *survival time* for the cerebrovascular syndrome is dose-dependent, but there appears to be a "threshold" dose-level, species-dependent, at and above which the syndrome comes into existence.

2. Clinical signs include agitation, apathy, disorientation, loss of equilib-

rium, loss of coordination of muscular movements, diarrhea, vomiting, tetanic spasms, convulsive seizures, coma, and death.

3. The principal changes noted are infiltration into the meninges, vasculitis of the brain, and edema. Neuronal cells of the cerebellum undergo pyknosis and nuclear shrinking indicating a disturbance in fluid balance in these cells.

4. Death is attributed to (although not proved to be caused by) neuronal damage due to vasculitis, edema, and increased intracranial pressure.

General Reference

Bond, V. P., Fliedner, T. M., and Archambeau, J. O.: Mammalian Radiation Lethality. New York, Academic Press, 1965.
McLean, A. S.: Early adverse effects of radiation. Br. Med. Bull., 29:69, 1973.

Chapter 15

EFFECTS ON DEVELOPING EMBRYOS

15.1 Introduction. The embryo probably represents the most radiosensitive stage in the life of any organism. This is true for several reasons: (1) Many of the cells in an embryo are differentiating, and, as has been implied in the law of Bergonié and Tribondeau, differentiating cells are quite radiosensitive. (2) There is, in the embryo, a high rate of mitotic activity; the system is *proliferative*, and, as has already been shown, mitosis is a sensitive time in a cell's life cycle. (3) An embryo is composed of "few" cells; many of its cells are ancestral to vast numbers of cells in the adult, so that damage done to one of them will be distributed over considerable numbers of descendant cells and over large areas of the postembryonic body. (4) If an embryonic cell is killed—at any time in the development of certain embryos and after certain stages of development of nearly all embryos—the cells for which it would have been an ancestor will not be formed (sometimes, however, another cell will or can take over its function). These factors, then, make the embryonic period an extraordinarily sensitive time of life with respect to radiation.

Within the embryo itself, as in adult organisms, not every organ or system will be equally radiosensitive. Unlike adults, however, in whom the same relative radiosensitivities always seem to exist (bone marrow, for example, is always about as radiosensitive as lymph nodes but more radiosensitive than the gastrointestinal system), there will be shifting radiosensitivities from one organ or system to another. A broad generalization can be made—the organ system *differentiating* will acquire great radiosensitivity and be profoundly affected by irradiation.

15.2 Regeneration and Repair. An embryo possesses, usually in much higher degrees than does an adult, the ability to regenerate lost parts. From the embryo's early stages, phagocytes are present which can resorb dead cells—those damaged and killed by irradiation. At early stages, with the dead cells gone, the remaining undifferentiated cells (undamaged ones) will begin to assume the role of the missing ones so that the *function* of the dead cells is not lost to the embryo. As a consequence, however, irradiated embryos are usually *smaller* than normal ones; the progeny of the deleted cells are never entirely replaced. The irradiated embryos *appear* grossly normal, but cytologic and histologic analyses show damage to their cells, their life span is reduced, and their *behavior* performance may not be equal to that of their unirradiated kin.

15.3 Stage of Development. Ionizing radiation is not the only agent that produces embryonic deformities or upsets in development. Nor are the upsets it produces characteristic of its own actions. Many agents (chemicals, viruses, other physical agents) produce abnormal embryonic development and the abnormalities they produce are identical with those caused by an exposure to ionizing radiation. The abnormality produced, in fact, is more characteristic of and dependent upon the system developing when the embryo-deforming agent is used rather than upon the agent itself. Differentiating or developing systems will be sensitive to or easily damaged by any number of agents. The reactions of the system (damage to the cells in it) will always be expressed in the same way— incomplete or abnormal development *of that system*. Radiation interactions are, of course, random and indiscriminate. If an entire embryo is irradiated, a differential reaction of its organ systems will be observed, based on which of these is in differentiation at the time. The *stage of development* of the embryo, then, will determine which organ system of the totally irradiated embryo will be most seriously damaged and, consequently, the kind of abnormality that will subsequently be observed. Chemical agents and even viruses *may* (although not necessarily) be more selective in the system that they will affect. If this is the case, the defect will probably be a reflection of nonrandom distribution in the tissues of the embryo rather than a specific mode of damaging action of the agent used.

15.4 The Developmental Abnormalities. In mammals, irradiation of the developing embryo before it becomes implanted into the uterus usually results in death rather than the production of abnormalities. The survivors ordinarily appear normal. Fertilization of the mammalian ovum most often occurs in the tubes leading from the ovary to the uterus. It takes place, then, in the oviducts and the very earliest embryonic development goes on in these tubes. Even after the cleaving and developing zygotes arrive *in* the uterus, they need not implant themselves in its wall immediately. In species (rats and mice) where implantation times have been determined, the zygotes may float around in the uterine lumen from several hours to days before implantation occurs. During this time the ovum continues to develop. In its preimplantation stages it is evidently very susceptible to the *lethal* action of radiation. It has been shown that, even with doses as low as 200 rads, up to 80 percent of early preimplantation embryos of mice are killed. In later preimplantation stages the embryos are less sensitive (for example, 40 percent are killed if irradiated on the fifth day).[1] Such results are reasonable, for the embryo in these early stages consists of very few cells and these will be in the process of differentiation. The cells will be sensitive to radiation, and a loss of a few of them will not be a trivial matter from the point of view of the entire embryo.

Irradiation of the embryo itself during organogenesis leads to malformations and to *neonatal* death (rather than, as when it is in the irradiated preimplantation state, *prenatal* death). In the mouse this period is from 7 to 12 days after fertilization, a period equivalent in human beings to the *second* through the *sixth* week of pregnancy. Within this period the defect produced depends on the system developing when irradiation occurs, but, in the mice, *microphthalmia* (tiny eyes) typically results after irradiation at day 7 to 8. *Colosteoma* (a fissure of the eye), *spina bifida occulta* (a defect in the closure of the spinal canal), *hernia*, and *urogenital anomalies* are seen after irradiation at days 9 and 10.

In rats, other workers have demonstrated *anencephaly* (the absence of cerebrum and cerebellum as well as the flat bones of the skull), *anophthalmia* (missing eye or eyes), and *hydrocephalus* (fluid in and around the brain), as well as other brain deformities, after irradiation on the ninth, tenth, eleventh, and twelfth days, respectively. *Microcephaly* (pinhead) is seen after irradiation from the twelfth day (corresponding, in man, to the fourth week) to birth.

Low doses (25 rads acute x-ray exposure) in the 7½ day mouse greatly increase the incidence of skeletal anomalies, and, in an inbred strain, doses of 18 rads given on the eighth day result in malformations in 25 percent of those irradiated.[2,3]

The greatest abundance of malformations occurs after irradiation *early* in development. During the first third of pregnancy the embryo is the most vulnerable, in the second third it is less so, and in the final third it is quite resistant. Male germ cells, however, become very sensitive during the latter period. In mice, acute x-ray exposure of 150 rads at the nineteenth day in gestation (it is a late fetus) results in sterilization. Female germ cells have a similar vulnerability, but one that occurs earlier—about the fifteenth day. Their sensitivity then declines until birth.

15.5 Human Experience.

Evidence for the production of anomalies in human embryos by radiation is too scanty to give an accurate picture of the situation. Different abnormalities have been attributed to irradiation of the embryo; of these, *microcephaly* and associated conditions with microcephaly are the best documented. Frequently, abnormalities of the central nervous system, the eye, and the skeleton are ascribed to irradiation, but the data are such that reliable prediction of dose related to effect and expected incidence is not possible. It can be said, however, that every anomaly of the human fetus has been produced in either mouse or rat by using x ray at comparable stages of development of the animal embryo. While this does not justify extrapolation or prediction of results in human beings, it does make *reasonable* the assumption that similar results might be expected in irradiated human embryos at comparable stages of development (this does, however, remain an assumption).

Many children exposed *in utero* at Hiroshima and Nagasaki were found to be mentally retarded and to have head circumferences less than those expected or well below the mean for their age (all were 7 to 15 weeks in gestation—a radiosensitive period, but not the most radiosensitive one if predictions made from mice are valid). All the children who were 7 to 15 weeks in gestation and whose mothers were less than 1200 meters from the point of detonation of the bombs (the highest dose of radiation is expected in this area) were affected in this way. Those in the same period of gestation but whose mothers were at a greater distance from the point of detonation were less affected. Of the 22 in this gestation period whose mothers were at 1201 to 1500 meters from point of detonation, only 18 percent were affected. These abnormalities were not noted in children exposed *in utero* whose mothers were farther than 1500 meters from the detonation point. Lesser abnormalities were also encountered in association with severe microcephaly; among them were strabismus, an abnormality of the eyes in which the visual axes do not meet at the desired objective point (it is a consequence of incoordinate action of extrinsic ocular muscles), congenital dislocation of the hips, and mental retardation. An increased incidence of congenital heart disease and hydrocele (an accumulation of fluid in the sac of the tunica vaginalis of the testes) was observed. Fetal death was greatly increased as compared with that in the unirradiated population. At Nagasaki, the rate of fetal deaths was 23.4 percent and that of neonatal or infant deaths was 26 percent in those under 2000 meters from the point of detonation. Among the infants who survived, 25 percent were mentally retarded. Overall, the morbidity and mortality in this group was 60 percent; in a similar group from the unirradiated Japanese population it was 6 percent.

Whether or not abnormalities are produced in embryos during radiodiagnosis is yet uncertain. Some instances of increased frequency of malformations of the eye have been reported, but they are not beyond dispute.

15.6 Clinical Implications. Whether exposures of ionizing radiation in the clinical dose range are embryo-deforming in human beings is still uncertain, but they may well have seriously deleterious effects. In experimental animals, embryos in the preimplantation stages are very sensitive to the killing action of radiation, and this situation *may* be similar in human beings. Studies in mice have shown that 100 percent of the embryos are malformed (many having multiple abnormalities) after exposure to 200 rads acute irradiation.[4] This has led to the recommendation that pelvic irradiation of women of child-bearing age should be restricted, as far as possible, to the 10 days following the onset of menstruation. This is a good recommendation, for not enough work has been done to allow good estimates of the effects of low doses given during early pregnancy. Such doses could

lead to death of the embryo (which can go unnoticed) or to a mutation involving very large numbers of cells. Later in pregnancy, irradiation carries a risk of somatic mutation; the most serious risk appears to be the induction of cancer (leukemia is induced).

Pelvic irradiation of women in whom there is a probability of pregnancy, particularly early pregnancy, should be avoided, except in serious emergency. Although isolated and too scanty to draw definitive conclusions, reports of malformations following clinical diagnostic exposure of the embryo to x rays have appeared. In mice, the administration of ^{131}I to pregnant females produces hypothyroidism in both the mother and her offspring. Tumors of the pituitary (adenomas) of the offspring may occur later. Isolated cases of similar destruction of human embryonic thyroid have been reported following administration of ^{131}I to the mother, but pituitary tumors have not as yet been known to ensue.

SUMMARY

1. Embryos are exceedingly radiosensitive. The sensitivity changes with age; generally, older embryos are less sensitive than young ones.
2. Critical times for producing radiation damage to various organ systems exist; these times probably coincide with the time the affected system is differentiating.
3. Human embryos are vulnerable to damage from radiation.
4. Irradiation of pregnant women should be restricted to necessary exposures; exposures of the pelvis should be limited, if possible, to emergency procedures.

Text References

1. Russell, L. B., and Russell, W. L.: Cold Spring Harbor Symposia on Quantitative Biology. *19*:50–59, 1954.

2. Russell, L. B., and Russell, W. L.: Radiation hazards to the embryo and fetus. Radiology, 58:369–377, 1952.
3. Russell, L. B.: Effects of low doses of x-rays on embryonic development in the mouse. Proc. Soc. Exp. Biol. Med., 95:174–178, 1957.
4. Miller, R. W.: Delayed effects occurring within the first decade after exposure of young individuals to the Hiroshima atomic bomb. Paediatrics, 18:1, 1956.

General References

International Commission on Radiological Protection: Committee I. The evaluation of risks from radiation. Health Phys., 12:238–302, 1966.
Ruge, R.: In Mechanisms in Radiobiology, Vol. II, Chapter 1. M. Errera and A. Forssberg (eds.). New York, Academic Press, 1960.
Russell, L. B.: In Radiation Biology, Vol. I, Part 2, Chapter 13, pp. 861–918. A. Hollaender (ed.). New York, McGraw-Hill, 1954.

Chapter 16

EFFECTS ON IMMUNITY

16.1 The Immune Mechanism. The natural defense mechanisms that protect an organism against harmful agents in the environment may be either innate or acquired. The innate defenses are a product of the genic makeup of the organism and are an expression of the experience of past generations. Acquired resistance is related to the experience of the individual with foreign substances (although heredity can modify this acquired resistance).

The basis of immunity is the antigen-antibody reaction. The immune response is initiated when *antigen* (i.e., infectious agents, tissue grafts, foreign proteins) enters the organism and stimulates the lymphoid-macrophage system. Under appropriate conditions the antigen is capable of inducing the formation of *antibodies* (specific proteins that have the capacity of neutralizing or reacting with the antigen). This specificity is a fundamental nature of immunity; the body reaction to a specific organism does not confer protection against another unrelated organism. For instance, after an attack of measles, humans are immune to further measles infections but not to other agents such as mumps or polio viruses.

The body thus "remembers" the initial challenge by the agent and is able to respond at a later time (acquired immunity).

The ability to recognize a particular antigen and to distinguish it from another is just a first step. The organism must also be able to differentiate between "self" and "nonself," that is, recognize what is foreign.

It would appear that circulating body components are able to reach the developing lymphoid system in the perinatal period and result in "learning." A permanent tolerance or unresponsiveness is created so that at maturity there is an inability to respond to self body components.

In acquired immunity, the specific response to an antigen is not the same for the first and second contacts. For instance, after injection of a toxoid into a rabbit, several days will elapse before antibodies are detected in the blood (Figure 16.1). The blood levels will reach a peak and then fall. If the rabbit is allowed to "rest" and then is given a second injection of the toxoid, the response is quite different. Within a few days the antibody levels rise steeply to reach much higher levels than

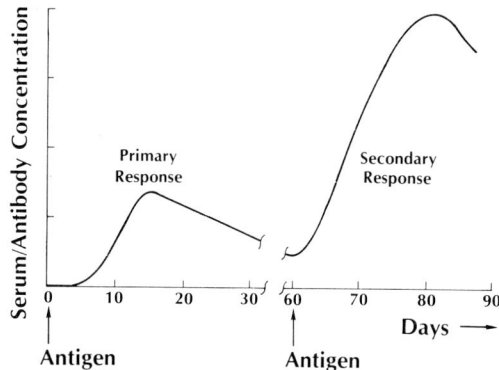

Figure 16.1. Primary and secondary responses. Animal injected with toxoid on two separate occasions. The antibody response on the second contact is more rapid and intense. (Redrawn from Roitt.)

those observed in the *primary response*. This *secondary response* is characterized by much more rapid response and higher antibody levels, since it results from "turning-on" the population of memory cells in the antibody-forming system which rapidly react to the second challenge by the antigen. Vaccination utilizes this principle by employing relatively harmless antigen (e.g., killed virus) as the primary stimulus. This imprints a memory which alerts the body defenses to subsequent challenge by the virulent form of the organism. The rapid secondary response usually prevents the infection of the virulent form from taking hold.

16.2   Cellular Antibody Production. When antigen enters the body, two types of reaction result: (1) *humoral antibody* and (2) cell-mediated reactions. The synthesis and ultimate release of free antibody into the blood and other body fluids result in direct combination with and neutralization of bacterial toxins, by coating bacteria to enhance phagocytosis. These *humoral antibody* activities are usually associated with the immunoglobulin fraction of the serum. In *cell-mediated* reactions, the production of sensitized lymphocytes that have the anti-

body on their surface is responsible for such reactions as the rejection of tissue transplants.

16.3   The Role of the Small Lymphocyte. In the resting lymph node, about 80 percent of the cells are small lymphocytes (about 7μ in diameter), and the remainder might be classified as large lymphocytes. Plasma cells are rare. After antigenic stimulation the number of plasma cells rises abruptly to large numbers. It would appear that the small lymphocyte is probably the cell type that responds initially to the antigen, by differentiating into a large, rapidly dividing cell called the immunoblast (Figure 16.2). It is further proposed that the immunoblasts divide asymmetrically into larger, antibody-synthesizing plasma cells and into smaller, resting lymphocytes that persist for long periods without further division as dormant "memory" cells for the specific antigen. Thus these small lymphocytes that carry the memory of the first contact with the antigen are responsible for the previously described secondary immune response.

16.4   The Role of the Thymus. Since the thymus gland is populated mainly by cells that are from appearance indistinguishable from lymphocytes, the thymus has been long suspected to play a role in the

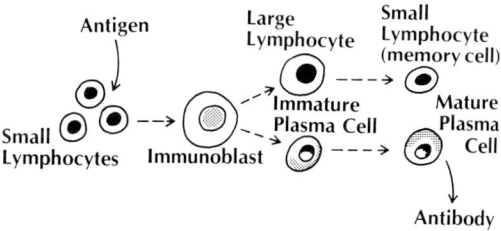

Figure 16.2. Possible sequence of changes from antigenic stimulation of the small lymphocyte in the immunologic response. Differentiation into lymphoblasts is followed by cell division and differentiation into long lived "memory" cells and plasma cells which produce most of the circulating antibodies. (Redrawn from Davis.)

immune mechanism. This idea was reinforced by the observation that many children with thymic abnormalities had immunologic deficiency disorders. Recent studies seem to indicate that the maturation of peripheral lymphoid tissues and their acquisition of immunologic competency in young animals are greatly influenced by the thymus and some other lymphoepithelial structures (the bursa of Fabricius in chickens, and gut-associated lymphoid tissue such as the tonsils, appendix, and Peyer's patches).

16.5 Irradiation and Immunity. The widespread interest and publicity associated with human organ transplantation and the associated use of radiation as an immunosuppressive agent have made this an area of special interest. Even the layman is aware that immunologic suppression with radiation extends the life of an organ graft, but that the recipient is also made more susceptible to infection. Because the lymphoid system is in the same range of radiosensitivity as the sensitive hematopoietic system, whole body irradiation at relatively high levels results in rapid death of large numbers of lymphocytes and inhibition of cell division in many others. Fractionated irradiation and radiation delivered at low dose-rates have a lesser effect on immune mechanisms than a single acute dose.

In mammals, doses of whole-body radiation of the order of 500 rads cause rapid degeneration of lymphoid cells. Active proliferation usually begins about 3 or 4 weeks later, with lymph nodes appearing normal shortly thereafter. If an antigen is introduced after irradiation, the formation of antibodies is delayed until the capacity for mitotic activity is recovered (remember that cell division and differentiation, Figure 16.2, are necessary for a primary response). If, however, the antigen has stimulated some cellular proliferation just before irradiation, the lymphoid cells can apparently continue their differentiation and eventually form antibodies. However, the time interval between introduction of antigen and appearance of antibody in the blood will be prolonged and the rate of rise in blood levels would be reduced. Cells that are actively synthesizing antibodies may continue to be active even after massive doses of irradiation. The secondary immune response is also relatively resistant to irradiation. The appearance of antibody may be delayed, but peak levels are usually normal. The formation of antibody may actually be unusually intense, even when compared to that in control animals, especially when the antigen is introduced before irradiation. This is called *enhancement* and is probably related to increased activity of stimulated memory cells in an environment of lymphoid tissue that has been depleted of many other cells.

16.6 Protection and Restoration of Antibody Formation. The depression of antibody formation by radiation can be modified by a variety of techniques that include shielding a portion of the lymphoid tissue of the body, the injection of living cells, and the use of nucleic acid derivatives. The protection by partial body shielding is related directly to the amount of lymphoid tissue protected. For instance, because the appendix contains many times the lymphoid tissue found in the spleen or legs, appendix-shielded animals recover sooner and produce more antibody than spleen- or leg-shielded animals (Figure 16.3). In adult animals given sufficient doses of irradiation to destroy their immunologic responsiveness, injection of bone marrow, spleen, and lymph node cells restored immunologic competence. The injection of nucleic acid degradation products at the same time as the antigen can restore the antibody-forming capacity after irradiation. Colchicine and endotoxins, cytotoxic agents, also appear to act in the same way

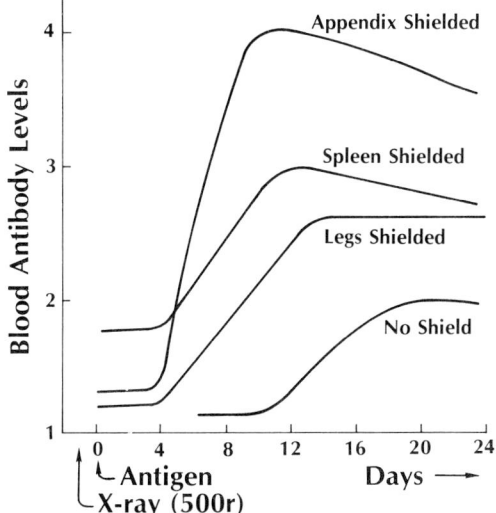

Figure 16.3. A comparison of partial body shielding to whole body irradiation. The end point is antibody (hemolysin) production in the rabbit. Note the progressive increase in protection afforded by shielding the legs, spleen, and appendix. (Redrawn from Taliaferro et al.)

by releasing nucleic acid products from injured cells.

16.7 Alteration in Immunity to Infection after Irradiation. The increased susceptibility to infection from nonpathogenic, as well as pathogenic, organisms plays a significant role in the survival of irradiated animals. Irradiated animals are capable of producing antibody even when irradiation occurs shortly after infection. However, the delay in production of antibody may lead to death in spite of their potential to produce antibody in normal amounts. The breakdown of innate immunity probably plays the most significant role. Within days after irradiation there is a rapid drop in the bactericidal activity of the serum. Large doses of radiation allow increased movement of intestinal flora through the wall of the intestine. Even if the bacteria are phagocytized in the mesenteric lymph nodes, the irradiated macrophages are unable to neutralize them. Within a short time they escape into the blood where they are phagocytized by the reticuloendothelial cells of the liver and spleen. Because these macrophages are also impotent, the bacteria are again released into the blood, resulting in an overwhelming septicemia and death.

16.8 Secondary Disease. As was mentioned earlier, the transplantation of bone marrow or spleen cells into lethally irradiated animals noticeably prolongs survival. The survival of the "graft" and the irradiated host is dependent upon the compatibility of the tissue of the host and the donor. The more closely related the donor and host, the more successful and longer lasting the transplant. The survival of homologous (same species but genetically different) or even heterologous (different species) transplants increases with increasing radiation dose. At low doses the chance of a "take" is reduced because the immune response of the host may quickly recover and rapidly reject the graft. At higher doses the immune response is deeply depressed or destroyed. Thus, at low doses, homologous or heterologous transplants of bone marrow will at best have no value and may even be toxic to the host.

If homologous bone marrow "takes" in an irradiated individual, it will be capable of providing needed blood cells and will allow the organism to survive the acute effects of irradiation. Later, however, the same organisms begin to die and, with time, die in increasing numbers. Death appears unconnected with radiation exposure. Rather it appears to be the result of a reaction of donor cells against the host (graft vs. host or GVH reaction). The response has also been called "secondary" or "runt" disease and is characterized by skin lesions, gastrointestinal dysfunction, atrophy of lymphoid tissue, and generalized "wasting" of the tissues. The intensity of the reaction and the time of onset depend on the genetic relationship of the host and the donor. The more closely the two are related, the longer before secondary disease begins and the slower its progress. In bone marrow transplantation the reaction is potentiated when small amounts of lymphocytic tissue are transfused with the marrow.

SUMMARY

1. The immune mechanism appears to be of two types: innate and acquired.
2. The immune response is an antigen-antibody reaction.
3. The reaction of the organism to an antigen entering the body is either humoral or cell-mediated.
4. The small lymphocytes and the thymus appear to have roles in the response to antigen.
5. Irradiation is immunosuppressive but makes the recipient more susceptible to infection and in some cases has enhanced the response.

General References

Davis, B. D., et al.: Microbiology. New York, Harper & Row, Chapter 15, pp. 454–508, 1970.

Jaroslow, B. N.: Radiation and the immune response. In Medical Radiation Biology, G. V. Dalrymple et al. (eds.). Philadelphia, W. B. Saunders Co., 1973.

Roitt, I. M.: Essential Immunology. Oxford, Blackwell Scientific Publications, 1971.

Taliaferro, W. H., Taliaferro, L. G., and Jaroslow, E. N.: Radiation and Immune Mechanisms. New York, Academic Press, 1964.

Chapter 17

TREATMENT OF IRRADIATED ORGANISMS

17.1 Introduction. The syndromes leading to death in lethally irradiated organisms and even the "sickness" following irradiation at nonlethal, total-body dose-levels have been ascribed to radiation damage to cells of *all* the tissues of the body. But death, after exposure to a wide range of dose-levels, has been presumed to result *primarily* from radiation damage to and failure of the cell renewal systems of two major organ systems, the hematopoietic and gastrointestinal systems. Accordingly, if this presumption is correct, a therapy or course of treatment for irradiated organisms could be devised which might make it possible to mitigate the symptoms of the syndromes and to measurably extend the life of irradiated organisms, hypothetically, for periods of many years.

17.2 Rationale. The treatments developed have been based upon examination of the primary and secondary consequences of total-body radiation. Lost or vanishing functions or cells, those that are destroyed by total-body exposure, are carefully *replaced* until such time as the irradiated organism can begin to regenerate its missing cells or functions.

This can be done;the principle is well established. Successful, long-term treatment, however, depends to a great extent on the total dose received. After exposure to very high dose-levels, the damage done to the central nervous system and the consequences that flow from it are, at present, beyond treatment. But, that category aside, even at lower doses the actual dose received will still determine the ultimate success of treatment. This is true, because dose is the determining factor in how long natural regenerative processes are delayed following radiation exposure. Replacement therapy, the replacement of cells or functions, cannot, at present, go on for protracted periods of time; successful treatment presupposes that the irradiated organism can soon again become self-sufficient.

17.3 Secondary Manifestations. The secondary manifestations that must be treated following total-body radiation exposure if even short-term survival is to be possible are granulocytopenia (the absence or scarcity of granulocytes—mature polymorphonucleocytes—from the circulating blood), thrombocytopenia (the absence or scarcity of thrombocytes—

platelets, red blood cells—from the circulating blood), and, in species where this occurs, anemia. At higher doses, the watery, often bloody diarrhea, which causes fluid loss and electrolyte imbalance, must also be controlled.

17.4 Susceptibility to Infection. Resistance to infections by nearly any kind of bacteria is markedly reduced following total-body radiation. Although it is true that this increased susceptibility may be due *in part* to a breakdown of the *barrier function* of various tissues (in particular that of gut after relatively high doses of radiation), a correlation can also be made between the time of onset of infection and depression of *granulocyte* count following radiation. Such a correlation suggests that these cells are an important part of the defense against infection.

Further, after total-body radiation, when granulocytopenia becomes detectable, replacement of granulocytes (by transfusion) clears up or mitigates the bacterial infections that follow lethal exposures.

Replacement therapy with granulocytes establishes an important point, even if its practical application gives, as yet, disappointingly poor results. Replacement does modify the progress and outcome of the syndrome. The poor results are due primarily to the rather short life expectancy of the transfused granulocytes. The half-life in man has been estimated at between 6 and 8 hours.[1,2] Such a brief survival time means that, for replacement therapy, a large supply of fresh granulocytes will have to be available. Repeated transfusions will have to be made during the period of granulocytic insufficiency (a period that will vary in length with dose). This in itself is a difficulty, for large supplies of these cells are not freely available, except from leukemic patients. In addition, repeated transfusions of granulocytes lead to a shortening of their already short life span

after transfusing. Nevertheless, none of these shortcomings of the treatment obscures the importance of the principle it establishes; replacement of granulocytes can modify the extent of infection and the course of the syndrome following lethal irradiation exposure.

17.5 Treatment of Infection with Drugs. Since the multiplication and proliferation of bacteria in the blood and the products that these bacteria produce are believed to be of major importance in radiation-produced lethality, it is reasonable to expect that antibiotics could substitute for the presence of granulocytes, prevent or mitigate infection, and extend life after lethal irradiation. This is the case; in the period of severe granulocyte depression, antibiotics do provide protection against radiation and forestall death. After exposure to radiation in low dose-ranges, antibiotics and platelets have been used in dogs and have prevented death. Antibiotic therapy is begun when signs indicate that infection is present, when the temperature becomes elevated. Unfortunately the infecting microorganisms in time become resistant to the antibiotic being used; the temperature again becomes elevated. But, if a different antibiotic is then given, the organisms can be controlled, and fever will again drop.

Equally successful therapy has not been reported for every study using antibiotics. In some irradiated species, some infections appear more susceptible than others. Nevertheless, the principle seems established; granulocytopenia can be treated either by transfusion of fresh granulocytes or by antibiotic therapy. Either alternative modifies the consequences of total-body radiation and can extend life beyond that expected after lethal doses of radiation.

17.6 Control of Hemorrhage. Even if granulocyte replacement therapy is successful and infection is avoided, the hazard of hemorrhage due to loss of thrombocytes remains. In the untreated or unmodified acute radiation syndrome, infection and bleeding occur so closely together that it is not possible to separate their individual roles in bringing about death. But, in granulocyte-treated irradiated organisms, those in which infection is controlled, the role of bleeding and the extent to which it may be responsible for radiation death can be evaluated. Acutely irradiated, infection-free organisms die as a result of their exposure to radiation. Moreover, in such animals, death coincides with the time at which thrombocytes become significantly depleted. When platelet levels drop very low, irradiated animals will die unless they are treated. If *fresh* platelets *are* transfused, however, bleeding dramatically stops,[3] and death is averted. Nakamura has demonstrated that transfusion of thrombocyte-rich plasma to irradiated mice increased 30-day survival from 42 to 100 percent at 708 rads, from 0 to 90 percent at 755 rads, and from 0 to 26 percent at 850 rads.[4] While an absolute causal relationship is not yet established, the strong inference implicit in such findings is that death would have been caused by *thrombocytopenia.*

The effect of *fresh* platelet transfusion is evidently linked to the fact that such transfusions increase the level of circulating platelets. No significant prevention of bleeding has been noted with the use of platelets treated in various ways: frozen, disrupted, or lyophilized. Platelets in any of these forms do not increase the levels of circulating platelets nor do they prevent bleeding. The storage and preservation of platelets are, in fact, important factors in limiting the usefulness of this technique for treatment of radiation exposure. *In vivo* survival time of transfused platelets is relatively long compared to that of granulocytes (5 days, but there are significant species variations), long enough,

in fact, so that a method of storing platelets would have considerable clinical significance. These cells, under normal storage conditions, quickly lose the qualities that are needed in the treatment of radiation exposure.

When repeated platelet transfusion is necessary, the efficiency of this therapy for stopping bleeding gradually decreases and the life of the transfused cells grows shorter with each succeeding transfusion. This appears to be due to the development of an immunologic response on the part of the host against the foreign cells. Antibodies are formed (although rather more slowly in totally irradiated hosts than in unirradiated hosts) which agglutinate and destroy the transfused cells.

17.7 Treatment of Dehydration. When, after relatively high doses of total-body radiation, the gut is denuded and there is severe fluid loss, if time enough is to be given for the mucosa to regenerate and reline the gut, the water loss must be treated so that early death does not occur. Fluid replacement combined with antibiotic therapy is useful; in dogs it is effective in the early acute phases. Later, when infection is likely to be present, it is less useful.

17.8 Bone-Marrow Replacement. The transfusion of bone-marrow cells to lethally irradiated individuals would be expected to confer significant beneficial therapeutic results. This supposition is based on the observation that, in animals irradiated to high dose-levels but in which the spleens were shielded, significant protection against the radiation was obtained (Section 16.6). Hematopoietic cells appeared to be able to repopulate the bone-marrow spaces, thereby averting bone-marrow depletion. This could be taken as presumptive evidence that if bone marrow were transplanted to lethally irradiated organisms, the transplanted marrow would multiply to fill the depleted marrow spaces, and, in time, begin to supply mature circulating cells for the irradiated organism.

The hypothesis has been tested in many places, and many times bone-marrow transfusions do indeed increase the probability of survival of lethally irradiated organisms. The degree of increase is dependent upon the amount of bone marrow transplanted. The more that is transplanted, the better the chance for survival. The transfused bone marrow begins to multiply and replaces the irradiated organism's lost marrow. The more marrow transfused, the more rapid and the earlier is this replacement and the shorter the period in which the irradiated organisms will have a shortage of mature circulating cells.

However, the probable success of bone-marrow transplants will vary in relation to the genetic relationship between the donor of the bone marrow and the irradiated host. Any tissue transplant will be more "successful" (persist longer in the host) the more closely the donor and the host are related. This is true, of course, for bone marrow. If, in the case of an irradiated host, the host's *own* bone marrow is used as the donor (this is called an autotransplant; it can be made only if bone marrow is removed before irradiation and then injected after irradiation), a permanent "take" can reasonably be expected; no complications are likely to result. If an exchange is made between *identical* twins, permanent transplants can also be expected. But, the more genetically separated donor and host become, the less successful are the transplants. If *homologous* bone marrow is used as a donor (the donor and host are of the same species but are not identical twins), much larger amounts of bone marrow must be transplanted to confer "protection" against the bone marrow phase of the syndrome than when *au-*

tologous bone marrow can be used. In addition, when homologous and *heterologous* transplants are compared (in heterologous transplants, the donor and host are *not* of the same species), more heterologous than homologous tissue is required to obtain the same result. Beyond that, the *length* of the extension of life (protection against the immediately lethal effects of irradiation) is also dependent upon the genetic relationship of donor and host. The more closely related the donor and host, the longer the survival (or, stated another way, the better the protection). If homologous marrow is used, it does protect against the immediately lethal effects; treated animals survive longer than untreated irradiated animals. Later, these same animals begin to die; the numbers that die greatly increase with time after irradiation. Death appears to be the result of an *immunologic* reaction—a tissue response—of the donor cells *against* the host (of the graft against the host). The intensity and time of onset of this "runt" disease will be determined by the genetic relationship of donor and host. The more closely the two are related, the longer it will be before the secondary sickness starts and the slower this sickness will progress.

SUMMARY

1. Substitution or replacement of lost cells or functions can protect against the damaging effects of irradiation.
2. Susceptibility to infection and bleeding can be prevented by granulocyte transfusion, by antibiotic therapy, and by fresh platelet transfusion. Control of fluid loss and electrolyte imbalance can be corrected by salt and fluid therapy.
3. Bone-marrow transfusions will repopulate depleted marrow and protect against death, but runt disease is a lethal, complicating factor.

Text References

1. Mauer, A. M., et al.: Leukokinetic studies II. A method for labeling granulocytes *in vitro* with radioactive diisopropylfluorophosphate (DFP32). J. Clin. Invest., 39:1481–1486, 1960.
2. Athens, J. W., et al.: Leukokinetic studies III. The distribution of granulocytes in the blood of normal subjects. J. Clin. Invest., 40:159–164, 1961.
3. Fliedner, T. M., et al.: Comparative effectiveness of fresh and lyophilized platelets in controlling irradiation hemorrhage in the rat. Proc. Soc. Exp. Biol. Med., 99:731, 1958.
4. Nakamura, W.: Effect of thrombocyte-rich plasma on x-ray induced bone marrow death in mice. Radiat. Res., 55:118, 1973.

General References

Bond, V. P., Fliedner, T. M., and Archambeau, J. O.: Mammalian Radiation Lethality. New York, Academic Press, 1965.
van Bekkum, D. W.: In Mechanisms in Radiobiology, Vol. II, M. Errera and A. Forssberg (eds.). Chapter 5, pp. 297–360. New York, Academic Press, 1960.

Chapter 18

THE LATE SOMATIC EFFECTS

18.1 Introduction. The effects of radiation, as already noted, fall rather naturally into two general categories: immediate or acute and late. Although *all* the biologic effects occur as a result of energy transfer to cells in the instant of irradiation, some effects are not expressed for a long interval of time after irradiation—in some cases for months or years. Those that fall into this latter category comprise the genetic effects and the late somatic effects. Genetic damage is not always *expressed* until mutant genes find themselves in a genetic environment that *permits* expression; this *can* require several generations of recombinations. In this chapter only the late somatic effects will be considered.

18.2 Modifying Factors. The nature of the late effects, indeed whether they will occur at all, is dependent on a number of factors. These have been considered separately in Chapter 10. Briefly, there will be the inevitable *species variations, dose-dependent responses (the radiation dose absorbed* will determine the frequency of occurrence of some of the late expressions of radiation damage), and *dose-rate* dependency.

18.3 Sterility. . Most data obtained on the induction of sterility following irradiation are taken from work done on experimental animals (many experimental studies on the effects of radiation on the sexual capacity and/or on the genetic material of man are immoral and, of course, unethical) so that the situation in man remains somewhat in doubt. Inferences may be made, based on animal studies, but data obtained in animals are not always extrapolatable to man so that a degree of uncertainty must remain.

It can, however, be stated that exposure to radiation in *males* does *not* appear to affect sexual capacity (potency or libido). Of course, exposure to radiation, even at dose levels well beneath lethal doses, does produce a kind of "sickness." This sickness, whose signs comprise nausea, loss of hair, loss of appetite, malaise, sore throat, petechia, diarrhea, and slight to severe weight loss, is—as is any illness—debilitating. There can be, as there can be during any illness, loss of sexual desire and responsiveness, but when the illness has passed (its acute phase is over) full restoration of potency and libido occurs.

Animal studies show, however, that while potency and libido are unaffected, *fertility* may be, at least temporarily, impaired. *Permanent* sterility can be produced if sufficiently high doses of radiation are given the gonads (permanent sterility would not be *likely* to result after a *total-body* exposure because the doses required to produce this result are so high as to bring about one of the radiation syndromes in the irradiated individual and cause death).

After a dose of radiation to the gonads (either as total-body exposure or as radiation of the gonads alone), there is a period of continued fertility, followed by a period in which fertility is impaired. Depending on dose, fertility may be impaired enough to have been lost altogether (*sterility*). This "sterile" period or period of impaired fertility is temporary; in time (the amount of time required is dose-dependent) fertility is restored. Sterility need not consist of a total loss of sperm; rather, to the contrary, normal viable sperm may even be present. Sterility may be produced if the *number* of sperm is sufficiently reduced so that the probability of fertilization of an egg is reduced to unlikely levels. Such sterility may be termed *functional* sterility.

As stated before, the situation in man is incompletely understood, but animal data indicate that within the dose range of 400 to 1000 rads (the precise dosages are dependent on species) a period of temporary sterility (not simply reduced fertility) is induced in males. At doses *below* 400 rads there is no clear-cut period of *complete* sterility. At doses above 1000 rads some reduction in fertility is seen even directly after irradiation—in the period where, at lower doses, normal fertility persists. Very high doses are required to induce permanent sterility.

These observed phenomena are usually explained on the basis of differences in radiosensitivity of cells of the germ line. The production of mature sperm begins in the testes with cells which are called *spermatogonia*. These are the undifferen-

tiated cells of the testis in which the *first meiotic* division occurs. This division results in two cells called *spermatocytes*. The spermatocytes undergo a *second meiotic* division, yielding four cells called *spermatids*. The spermatids then undergo a complicated cytoplasmic reorganization (the process is known as *spermiogenesis*), to give rise to mature sperm, but no further *divisions* occur. The rearrangement of cytoplasm or spermiogenesis adapts the sperm to the mobile function it must have in order to "seek" and fertilize an ovum.

Of the cells involved in the various stages in spermatogenesis, the spermatogonial cells are the most radiosensitive. Spermatocytes, spermatids, and even the sperm themselves are rather radioresistant. During irradiation, then, with dose-levels to the total body in the nonlethal range (400 rads or less) the spermatogonia are affected; some will be killed, and, in others, division will be inhibited. There will be a *reduction* in the number of these cells and, in time and as a consequence of the damage to the precursor spermatogonia, reduction in the number of cells formed during the other stages of spermatogenesis. After a time, inevitably, the number of mature sperm will be reduced, and, if this reduction is severe enough, fertility may be impaired or even sterility may occur. Directly after irradiation then, no change in fertility is expected (none occurs) because the mature sperm are not affected. Similarly, the cells of the intermediate stages (spermatids and spermatocytes) are not unduly damaged and they continue to mature. The supply of mature sperm will only be interrupted when a deficit occurs as a result of the damage done, at irradiation, to the spermatogonia. If the radiation dose is not too high, the dead or the damaged spermatogonia will be replaced and the proper number of mature sperm will eventually be restored. Thus, a period of fertility follows irradiation, succeeded by a period of impaired fertility, succeeded in time by a restoration of normal fertility.

It should be stressed that this discussion centers only around fertility, i.e., the capacity of a spermatozoon to *fertilize* an ovum (to enter and begin development of an embryo within an ovum). Spermatozoa are, with respect to fertility, radioresistant. However, genetic mutations may have been produced in the irradiated sperm.

In human beings, as has been stated, few data are available, but the information that does exist indicates that a *single* dose of 400 to 600 rads to the testes can or is likely to produce complete sterility. Temporary sterility follows lower doses; a *single* dose of about 250 rads brings about sterility for one year. Very much lower doses (as low as 30 rads), however, can reduce the sperm count and impair fertility.

Even the complete loss of fertility is without effect on libido, hormone balance, or on somatic cells of the testes. No impairment of the sex drive or of sexual capacity has been noted.

In *females*, in terms of the results of exposure of the ovaries to radiation, the situation is similar to that in males. The predominant effect of radiation on female fertility is a *shortening* of the *breeding period*. The permanently sterilizing dose varies with age; older females require smaller doses to produce sterility than do younger ones. In mice, a 150 rads dose of x ray is permanently sterilizing. There is, as with males, a fertile period following irradiation; sterility sets in at a later time. Again, as with males, mature germ cells (in this case, ova) are not radiosensitive and are capable of fertilization or union with the male gamete so that sterility would not be expected to come into existence until the mature radioresistant form becomes depleted and no new eggs mature (because the pool of radiosensitive oogonia have been destroyed).

The dose (total-body dose) required to produce sterility in human females is rarely received; it will be high enough (300 rads or more) to initiate a radiation syndrome. Radiation given directly to the ovaries (either as therapy to the ovaries themselves or when the ovaries are included in the field of radiation) can, of course, produce sterility. In this case, unlike that of males, radiation sterilization does have important effects on sexual capacity. This is true because ovarian production of hormones is closely related to the development, maturation, and discharge of ova. If these processes are halted (either by radiation or by any other mechanism), an "induced menopause" is the result, with the expected concomitants of natural menopause: diminished libido, "hot flashes," and in some individuals, depression, and loss or absence of menstruation.

18.4 Shortening of Life Span. As has been mentioned earlier, exposure to radiation over the total body shortens the life span. Radiation is also a carcinogen (a cancer-inducing agent), and it is to be anticipated that irradiated animals will more often develop cancers than will nonirradiated ones (Section 18.7). This is the case, and, in consequence, the life of irradiated animals will be expected (on the whole) to be shortened, since the exposure induces a fatal disease. But, even if all deaths due to malignant disease are excluded, the life spans of irradiated animals are shorter by statistically significant amounts than those of unirradiated ones.

The amount of shortening of life of small animals after total-body irradiation appears to be dose-dependent. Smaller amounts of radiation shorten life to a lesser degree than do large amounts. Over a range of doses the response has been shown to be both linear (as in Figure 18.1) as well as curvilinear. But it is not known whether any dose of radiation, however small, will shorten life. Some argue and

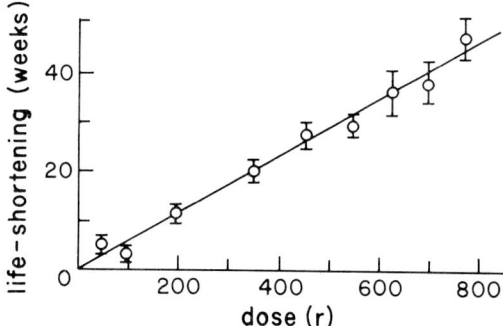

Figure 18.1. Life shortening as a function of dose for mice of both sexes. (From Lindop, P. J., and Rotblat, J.: Proc. Roy. Soc. (London), Ser. B, *154*:344, 1961.)

have argued that a "threshold" dose of radiation which is necessary to initiate the life-shortening effect may exist. Although no such threshold has been demonstrated, neither has it been shown conclusively that extremely small doses result in the shortening of life (experiments have shown no apparent threshold for life shortening within the experimental error). It is an exceedingly difficult point to test effectively, for very small doses will be expected (if the relationship is indeed linear *throughout*) to shorten life very little. Irradiated *populations* must be compared to nonirradiated populations to determine whether, on the average, the nonirradiated ones have longer life spans than do those who were irradiated. When only a small amount of time is involved, *very* large populations (under rather carefully controlled environments) must be used. Thus, any radiation-induced *disease* state (like cancer) which would shorten life must be excluded so that, in both groups, only animals dying from "old age" are compared. Experiments such as these are difficult to set up and run; their results are often even more difficult to evaluate.

Life shortening as a radiation lesion is subject to modifying influences. It is more noticeable if young or very old animals are irradiated; the presence of oxygen at irradiation enhances the effect; chemical protectors present during irradiation protect against the effect.

Life shortening is not restricted to the aftermath of *total-body* radiation; if only parts of the body are exposed, this effect can be elicited as well. The length of shortening, however, is less than that which follows total-body irradiation. While that is true, if the length of shortening is viewed as a function of the fraction of the body irradiated, irradiation of parts of the body produces a *relatively* greater effect than does total-body radiation.

18.5 The Effect of Dose-Rate. The life-shortening effect is dependent on *dose-rate* as well as on total dose. If a given dose is prolonged over a long period, its life-shortening effect will be much less than that of the same dose given in a short burst. If the dose is given in fractions, i.e., at interrupted intervals, the life-shortening effect of that dose will be greatly reduced, compared to the same dose given at one time.

18.6 Human Experience. The bulk of the data on life shortening in human beings is derived from studies made on the life spans of American radiologists. The life of American radiologists has, at least in the past, been shown to be shorter than that of other physicians. The observed shortening is believed related to day-by-day radiation exposure obtained in the course of the radiologist's work. At first, the suggestion that radiologists might have a shortened life was greeted with criticism, and the data upon which the conclusion was based were said by some to have too many shortcomings to demonstrate that point conclusively. However, at the same time, and over the years, American radiologists have become more and more careful to expose themselves to progressively smaller and smaller amounts of radiation. The data relating life span and radiation exposure of United States radiologists appear to have been substantiated. There has been demonstrable life shortening among radiologists in the past. But, with each decrease in radiation exposure received, the degree of shortening has been less. At present, and since 1960, no demonstrable life shortening can be detected.

18.7 Cancer. Ionizing radiations are carcinogens (cancer formers); exposure to ionizing radiation carries the risk of cancer induction in the irradiated organism. Ionizing radiation is a *general* carcinogen, that is, it induces or brings about cancers of any tissue in nearly any animal tested, irrespective of species.

Radiation is not the only carcinogen. There are, in fact, a multitude of them. Generally, carcinogens may be classified as radiations (both ionizing and nonionizing), chemicals (polycyclic hydrocarbons, azo dyes, to name two classes of compounds), physical chronic irritants (abrasives, cellophane implanted in an organ), and living agents (viruses). Among these, however, only radiation is at present known to be so general a carcinogen. Chemicals may induce a tumor in one tissue, but not in another; they may be carcinogenic in one species, but not in another. Viruses can be carcinogenic for certain species or strains or in certain tissues.

18.8 Latent Period. Following the use of radiation as a carcinogen, there is (as there is with chemical carcinogens) a *latent period*. It is a period in which the tissue that has received the carcinogenic stimulus does not appear abnormal. But, for radiation, at least, the carcinogenic transformations must have been accomplished in the moment of irradiation. They are not apparent and cannot be detected then—indeed, often not until a long time later; but they must essentially have occurred immediately when the

tissue was irradiated. What is going on during the latent period cannot now be described, for nothing can be detected. It is possible that, during the latent period, changes in irradiated cells are so subtle as to be beyond the present ability to detect them. On the other hand, it is also possible that the induction of cancer is a two-step process. The first of these might be the *initiation* of a cancer, the *transformation* of normal cells to cells having the capacity to become cancerous. But, a *second* stimulus might be necessary to actually start the *growth* of a cancer. The latent period might be a period of quiescence, the period between the two phases, *initiation* and *promotion* of growth.

Whatever the case, there is usually a long latent period (long relative to the length of life of the irradiated animal) between irradiation and the appearance of a tumor. In human beings the period can be 10, 20 or even 30 years. In addition to a long latent period, not *every* irradiated individual, even heavily irradiated individuals, will get a cancer.

Repeated doses of radiation will have the effect of shortening the latent period. In spite of the fact that there is such a long latent period, a cause-and-effect relationship between exposure to ionizing radiation and later cancer formation can be established. The major link in establishing this relationship is that the tumor almost invariably appears at the irradiated site (when local irradiation of a part of the body has been done) or in the site of localization of a radionuclide.

18.9 Leukemia. Only certain types of leukemias (acute leukemias and chronic myeloid leukemias) are known to be induced by ionizing radiation in man. Leukemia is observed after a short-term exposure over the total body or nearly all the body. It is also observed after external radiation in large amounts given at rather low dose-rates over the total body or to large parts of the bone marrow, and after an administration of radioiodine (^{131}I) in doses *greater* than *one curie* for therapy for carcinoma of the thyroid gland.

Leukemia is *suspected* (but not solidly established) to occur after deposition in the body of radionuclides such as radium, thorium, or strontium 90; in children exposed to diagnostic pelvic x-radiation *in utero;* after short-term exposures to doses of less than 100 rads; and in children receiving irradiation to the thymus gland early in their lives.

18.10   Relationship of Leukemia to Dose. The relationship of induced leukemia to dose among the survivors of Hiroshima and Nagasaki has been extensively studied, since they represent the largest group of irradiated human beings.

The precise estimation of the dose received and its composition (relative amounts of neutrons and gamma rays) are difficult to determine. The results from the Nagasaki experience have larger sampling variations, and the relationship of leukemia incidence to dose is unclear; however, the Hiroshima experience indicates a significant increase in leukemia for doses as low as 20 to 50 rads, and that the RBE is higher for neutrons, as compared to gamma rays, for leukemogenesis. The Hiroshima dose response appears to be "relatively linear" throughout, and, based upon this, 1 rad over the total body would produce about 3 cases of leukemia per million population per year. The annual incidence in these groups has been declining since the peak (5 to 7 years after exposure) but has not returned to control levels 20 years later. These data also seem to indicate that the radiation leukemogenic effect is stronger in males than in females and is greater in younger than in older survivors.

Studies have been conducted also in 14,554 patients whose spines were irradiated for the bone disorder, ankylosing spondylitis. These studies were also beset with difficulties, although not of the same nature as those encountered at Hiroshima and Nagasaki. The dose in the spine irradiations is known quite accurately, but doses to the same individual were given at different times, so that the total had to be obtained by addition. This is a serious shortcoming because the bone marrow is not likely to have been equally susceptible at all times. However, if this difficulty is ignored, a linear relationship leads to an expectation that 1 rad of whole body radiation would produce between one and two cases of leukemia per million persons irradiated.

In both therapeutic irradiation and irradiation as a result of the atomic explosions, the risk of leukemia as a function of time after irradiation diminishes (the longer time that passes after irradiation the smaller is the probability that any given irradiated individual will contract *radiation-induced* leukemia), but the rate remains higher than is encountered in similar nonirradiated environments for long periods of time.

It must be remembered that these values are *extrapolated* values; they are based on a *presumed* linear relationship through all doses. As yet, this has not been proved to be the case and no *proof* exists that the very lowest doses *do* induce leukemia in adults.

In the fetus there is evidence that doses in the range of 1 to 5 rads induce leukemia, but in this respect the fetus is believed very much more sensitive than is the postnatal individual.

A progressive rise in leukemia mortality rates over approximately the last 40 years has been described in the United States, England, and elsewhere. It has been suggested that increasing exposure to medical x rays is among the reasons for such an increase. In 1956, reports on the biologic effects of ionizing radiation were widely publicized. A decline in leukemia mortality rates has been observed; it may be related to awareness of the potential hazards of medical radiation.

Radiation-induced leukemia has been considered by some an occupational hazard for radiologists of the United States. A study encompassing the years 1948 to 1961 reported 12 cases of leukemia among radiologists, whereas the expected incidence for the population at large would be 4 cases. During this period, however, the frequency of leukemia among radiologists declined substantially (in the decade 1949–1958 the rate was half that of the years before 1949). The total number of cases is really quite small. Moreover, since the rate declined during this period, it would appear that safe practices are reducing the increased risk to a negligible quantity.

18.11 Radiation Leukemogenesis in Mice. The mouse has been used for most animal studies of radiation leukemogenesis because of the ease with which leukemia can be induced and the variety of leukemias associated with various inbred strains. In the mouse, granulocytic leukemia is less frequently induced by irradiation than lymphocytic leukemia (the opposite of man). The most widely studied form is lymphocytic leukemia arising in the thymus. Radiation-induced leukemia in mice appears to be influenced by a variety of factors such as strain, age, sex, and the presence of thymus. For instance, removal of the thymus prior to irradiation prevented induction of lymphoma in mice; removal of the spleen lowers their susceptibility to the induction of granulocytic leukemia. When unirradiated thymus is transplanted into irradiated thymectomized mice, it causes lymphoma in the graft. This "indirect" action of the radiation has led to speculation about the role of a humoral (or possibly viral) agent in radiation leukemogenesis.

18.12 Radiation Tumorigenesis. The production of tumors in irradiated animals and man was the first delayed effect of x

rays observed. Because the first x rays were only weakly penetrating, the first malignancies were confined to the skin. However, as more penetrating radiations came into general use, it became apparent that enough radiation to almost any body part increased the incidence of malignancy.

Skin Cancers. Large numbers of skin cancers occurred in early radiologists and patients who were exposed to relatively high doses were mainly carcinomas (malignancies of epithelial origin). In mice and rats, however, the most common radiation-induced skin tumors are sarcomas (of connective tissue origin).

Thyroid Cancers. A definite relationship between early childhood irradiation and subsequent thyroid cancer appears to be well established. Doses as low as 200 rads given for "thymic enlargement" proved to be carcinogenic. There is a lack of such a convincing association in adults. Thyroid cancers have been induced in rats by both x rays and radioactive iodine, but the role of hormones (such as TSH) and their relation to goiter-producing agents and iodine-deficient diets make interpretation of the details of the voluminous experimental data somewhat confusing.

Lung Cancer. Radium miners, who as a result of their occupations inhale quantities of radioactive dusts and gases (radon), often develop lung cancer. The induction of these tumors probably involves other factors such as irritation from dust as well as cocarcinogenic action of metals (arsenic and cobalt) in the mine dust. Lung tumors have been induced in rats and mice with massive single external doses or from locally deposited or implanted radionuclides in their lungs.

Bone Tumors. Painters who were employed in factories producing luminous dial watches and clocks in the 1920's had a high incidence of bone tumors. It was common practice at that time to bring the paint brush to a point by shaping it with their lips before dipping it into the luminous compound which contained long-lived radioactive elements (radium and mesothorium). In addition, during the same period, large numbers of patients were "treated" with oral or injected radium compounds for a variety of ailments. Many of these patients have since developed bone tumors. Tumorigenic effects of bone-seeking radionuclides other than radium have been documented only in laboratory animals. Doses of external radiation as low as 250 rads of x or gamma rays have been reported to induce osteogenic sarcomas in laboratory animals. Osteogenic sarcomas have also been induced in laboratory animals by the administration of both long half-lived (radium-226 and strontium-90) and short-lived (radium-224) alpha-emitting radionuclides which are bone seekers.

Endocrine Gland Tumors. A large variety of endocrine gland tumors have been induced in experimental animals; of these radiation-induced tumors, tumors of the ovary, pituitary, and mammary gland deserve special mention. The radiosensitive mouse ovary is also quite sensitive to ovarian tumor induction. For instance, in the RF mouse doses of x rays less than 50 rads are consistently tumorigenic. Spontaneous ovarian tumors are rare in mice; radiogenic tumors such as these have not been seen in other species. Pituitary tumors in mice and rats have been produced by destruction of the thyroid gland and by whole-body irradiation. Mammary tumors in mice and rats are influenced by genetic and hormonal factors and by the apparent presence of a virus-like agent that is specific for mammary tissue. Doses as low as 400 to 500 rads are capable of inducing mammary tumors in most female rats of certain strains. In males, the incidence of these radiogenic tumors is lower, and the latent period between exposure and appearance of the tumor is longer.

18.13 Mechanism of Action in Radiation-Induced Cancer. The etiologic basis of radiation-induced cancer (the mode of action that brings it about) is unknown. As with chemicals, which are other physical agents, the dose required and the mode of action are not clear yet. Deductions based on knowledge of cancers on cells and of normal tissues from which they are derived can be made concerning these points. There is much evidence to indicate that cancers arise as a result of *genetic mutations.* The indications are that (1) cancers seem to arise from normal tissue. Unless cancer cells which begin to grow only upon a stimulus are always present in normal tissues—and this is possible—normal tissues must be transformed to cancer tissue. (2) Cancers "breed true"; cancer cells give rise to other cancer cells, like themselves. (3) When transplanted from one part of the body to another (this process is called metastasis) or from one organism to another, cancer cells continue to breed true. (4) Cancer cells *retain* some characteristics of their tissue of origin.

It may be, then, that radiation produces a mutation in cells, one which frees the cells from growth control by the body in which the radiation occurs. This process would be the *initiation,* the transformation to a cancer cell. Then, if the transformed cell received a stimulus to grow and divide (freedom from the control of growth need not mean that growth itself *inevitably* follows), it will do so without restraint.

Not all workers accept the mutation hypothesis for the induction of cancer, and, in fact, this hypothesis has not been proved. Yet, the carcinogens as a whole (there are exceptions) *are* mutagens so that this process is, at least, quite possible.

18.14 Cataracts. Cataracts may form as a result of an exposure to ionizing radiation in which the eye is involved. They are relatively uncommon late effects of total-body radiation, for a rather large dose is required to induce their formation. The cataracts are not the same as those occurring as a result of senility; in appearance and development, they are distinct. Neutrons, however, do produce them in good quantity after rather low doses.

SUMMARY

1. The "late effects" following total-body exposure in low-dose ranges are impaired fertility, shortening of life span, cancer induction, and the induction of cataracts.
2. Impairment of fertility occurs because of radiosensitivity of precursor cells to the gamete.
3. Shortening of life span is a true radiation effect, but its genesis is not known.
4. Exposure to radiation increases the frequency of all kinds of cancer.
5. Cataracts occur after irradiation of the eyes, in particular with neutrons. It is not a common consequence of total-body radiation.

General References

Bacq, Z. M., and Alexander, P.: Delayed effects. In Fundamentals of Radiology, 2nd Ed., Chapter 17. New York, Pergamon Press, 1971.

International Commission on Radiological Protection: Committee I: The evaluation of risks from radiation. Health Phys., 12:737, 1966.

Ishimaru, T., et al.: Leukemia in atomic bomb survivors, Hiroshima and Nagasaki, October 1, 1950–September 30, 1966. Radiat. Res., 45:216, 1971.

Mole, R. H.: Late effects of radiation: carcinogenesis. Br. Med. Bull., 29:78, 1973.

Report of the United Nations Scientific Committee: The Effects of Atomic Radiation. New York, The United Nations, 1962.

Russell, W. L.: Genetic effects of radiation in mammals. In Radiation Biology, Vol. I, Part 2, Chapter 12, A. Hollaender, ed. New York, McGraw-Hill, 1954.

Stewart, A., and Hewitt, D.: Leukaemia incidence in children in relation to radiation exposure in early life. In Current Topics in Radiation Research, Vol. I, Part VI, M. Ebert and A. Howard, eds. Amsterdam, American Elsevier Publishing Company, 1970.

Van Cleave, C. D.: Late Somatic Effects of Ionizing Radiation. TID-24310, USAEC, 1968.

Warren, S.: The basis for the limits on whole-body exposure—experience of radiologists. Health Phys., 12:737, 1966.

Appendix A

MEASUREMENT OF RADIATION DOSE

A.1 Introduction. When dealing with the biologic effects of radiation, one is interested in the amount of energy absorbed by the biologic material rather than the energy that passes through it. With radiations such as x or gamma rays, an organism may absorb all or only a part of the incident energy, but in most instances absorbed energy will only be some fraction of the exposure. Although techniques based upon calorimetry and chemical reactions can measure absorbed energy directly, most often the absorbed dose is not directly determined. Instead, absorbed dose can be estimated or computed from the exposure (or other dosimeter readings) using certain assumptions about absorption of the radiation in the medium. What follows is a discussion of the techniques used to measure dose with emphasis on the most important technique, air ionization.

A.2 Calorimetric Methods. From an overall point of view, the most satisfactory type of dosimeter would be one that would directly measure absorbed dose or absorbed energy per gram of material irradiated. For most absorbing materials all but a small percentage of the energy absorbed from radiation ultimately appears in the form of *heat*. Thus a calorimeter which measures heat directly is, in principle, a very direct approach to the measurement of dose in rads. But it requires elaborate, refined equipment in its operation.

Calorimetric systems operate by absorbing energy from the radiation field, retaining the energy until it is degraded to thermal energies, and measuring the temperature rise of the system. The major problem in their use is the small amount of heat actually produced by even high doses of radiation. Acute lethal effects in mammals are seen with whole body doses of 500 rads. A simple calculation shows that a dose of this magnitude would produce a temperature rise of only $1/1000°$ C in water.

Today, sensitive calorimeters capable of precisely detecting minute rises in temperature are in use. They usually consist of a small thermally insulated mass of material centered in an extended medium. Thermal sensors to measure the temperature change and electrical heaters for calibration are included. The loss of

117

absorbed energy to forms other than heat (chemical energy) is a fundamental difficulty.

A.3 Chemical Dosimetry. Chemical dosimetry provides a practical method of dose measurement based on the quantitative oxidation and reduction reactions taking place upon irradiation of certain chemical systems when the quantity of the reaction is directly proportional to the radiation dose. Bacq and Alexander have enumerated four practical criteria which must be satisfied in such a system.[1] The reaction

(1) must be independent of the dose-rate over a wide range;

(2) must be carried out in dilute solutions such as water or benzene where the absorption is largely independent of photon wavelength;

(3) must be little influenced by changes in ion density over a wide range; and

(4) must be relatively insensitive to the presence of impurities.

Of all chemical systems available for routine use, the ferrous sulfate system is the most reliable and the most widely used. The ferrous sulfate or Fricke dosimeter (named after the chemist who first described the reaction) usually is made up of a dilute solution containing ferrous sulfate, sulfuric acid, and sodium chloride (which counteracts the effect of organic impurities in the water).

In the radiation reaction, ferric ions are produced from the ferrous ions. After irradiation, the amount of ferric ions produced may be determined directly in an ultraviolet spectrophotometer; the yield of ferric ions depends directly on the dose absorbed. The reaction proceeds only in the presence of dissolved oxygen. The amount of oxygen normally present in an aerated solution is usually completely exhausted after about 50,000 rads, but this upper limit may be extended if oxygen is supplied. Since the G value (the number of molecules changed per 100 eV in a chemical reaction) is quite precisely known, this aerated solution makes an excellent dosimeter because it is a tissue equivalent absorber (water) and the absorbed dose in rads may be quite precisely calculated. Its G value is quite constant for beta particles and x and gamma rays in the energies normally employed in radiobiologic and medical experimentation, but the yield does fall off as the ionization density (LET) is increased. A limitation of this system at the present time is that it requires doses greater than 4000 rads before an accuracy of better than a few percentage points may be achieved.

A.4 Radiation-Induced Thermoluminescence. When certain crystalline solid materials are exposed to ionizing radiations, the process of ionization produces free electrons that are subsequently trapped in imperfections in the crystal lattice. At normal room temperatures the electrons remain trapped on a relatively permanent basis. Upon heating, thermal agitation will release them, and they will recombine with opposite-charge carriers and emit light in the process. The light emitted is directly related to the absorbed radiation dose. Lithium fluoride (LiF), the most widely used, is little affected by the atmosphere, stores the signal with little fading, and, because of its low atomic number, is nearly air and tissue equivalent. Hence, for identical exposures to radiation, the amount of energy absorbed by LiF is very close to that absorbed by an equal mass of soft tissue. Because the response of LiF to radiation is linear between one roentgen and a few hundred roentgens, it has wide application in radiobiologic experiments.

A.5 The Measurement of Exposure. The measurement of exposure to x or gamma rays is of importance because the absorbed dose at a point depends upon the exposure and because absorbed dose values can be determined from exposure in roentgens at that point. Exposure measurements rely primarily on the measurement of the ionization of gases by radiation. Approximately one year after the discovery of x rays by Wilhelm Roentgen, Perrin, using a charged condenser with air between the plates, demonstrated a loss of charge caused by ionization of the air by x rays. In 1908 Villard proposed a unit of x ray quantity based on the ionization of air.

Air ionization is now the recognized technique for the measurement of exposure. It is rather easily accomplished, reproducible, and controlled. Air is of almost uniform consistency everywhere, and the electricity produced in it by radiation (ionizations) is relatively easily and accurately measured (Figure A.1).

Since air has an effective atomic number very nearly the same as the soft tissues of the body, the absorption of x-ray energy per gram by soft tissue, water, and air is almost the same, even when x rays of widely varying wavelengths are utilized.

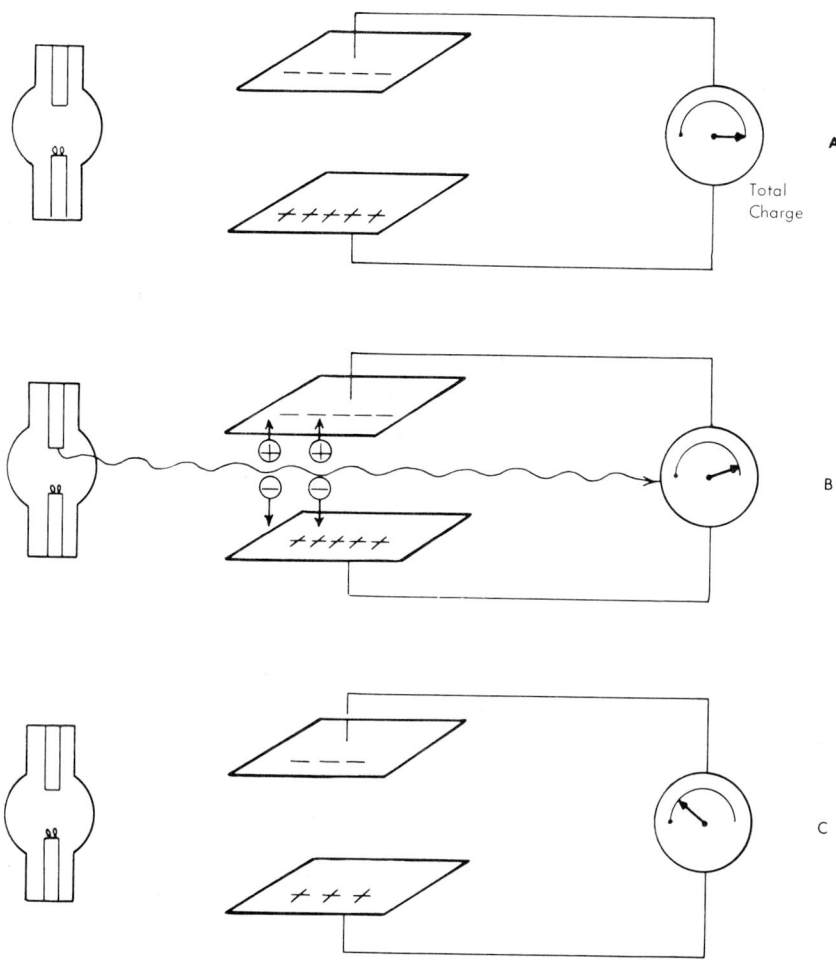

Figure A.1. Measurement of air ionization. A. A parallel plate condenser is charged. B. The air between the plates is irradiated by x rays and ion pairs (positive atoms and negative electrons are produced). C. These ion pairs migrate to the charged plates and discharge them. The radiation exposure is proportional to the measured discharge.

This factor makes it possible, with modifications, to extend measurements made in air to absorbed dose in tissue.

A.6 The Roentgen. The Second International Congress of Radiology (in 1928) adopted the roentgen as the official unit of radiation quantity. Over the intervening years the wording of the definition evolved to "The roentgen shall be the quantity of x or gamma radiation such that the associated corpuscular emission per 0.001293 gram of air produces, in air, ions carrying 1 electrostatic unit of quantity of electricity of either sign." Actually, 0.001293 gram of air is 1 cc of air at standard conditions of 0° C and 760 mm of mercury pressure. The "associated corpuscular emissions" consist of electrons ejected from the air atoms by the x or gamma ray photon-produced air ionization formed in 1 cc of air at standard conditions. In 1962 it was recommended that the roentgen be expressed in the International System of Units, 1 roentgen $= 2.58 \times 10^{-4}$ coulombs/kilogram of air. The roentgen, by definition, is a unit of exposure for x or gamma rays only; it cannot be used for other types of radiation.

Those secondary corpuscular radiations that originate within, but lose a portion of their energy outside, the specified volume are balanced by those that originate outside the volume, but contribute ionizations to it. Thus for practical purposes we may consider that all the ions produced within the volume lose all their energy within it. This is true only when the air volume extends in all directions at least a distance equal to or greater than the maximum range of any secondary electron produced by a photon of a specific energy. This includes, therefore, the distance from the source of radiation to the volume in question. This point of maximum ionization occurs at a distance from the radiation source equal to the maximum range of a secondary electron in the medium (in this case, air). This distance may be considerable; for 200 KVp x ray it is about 40 cm of air, and for 1000 KVp x ray it is about 3 m.

A.7 Measurement of the Roentgen. In order to measure exposure in roentgens, a known mass of air must be segregated, and the ionization produced within this air volume must be measured. This is accomplished by the use of a device called a standard free-air chamber, which consists of two parallel charged metal plates similar to those shown in Figure A.1. A thin x-ray beam is passed between the plates, and all the secondary electrons are collected.

Standard free-air chambers are not routinely used in most laboratories but are usually confined to institutions such as the National Bureau of Standards in Washington, D.C. They are regarded as primary standards, capable of measuring the roentgen. The thimble chamber is regarded as a secondary standard, calibrated from readings made with the primary free-air chamber, and provides a practical method of measuring exposure in the field and in the laboratory.

A.8 Electronic Equilibrium. An important concept associated with the measurement of exposure in roentgens is electronic equilibrium. It can be shown that as the energy of an x-ray beam is increased, maximum ionization occurs at *increasingly greater distances* from the radiation source. At distances beyond a point of maximum ionization, the measured dose is reduced because of air attenuation. When the number of electrons which are set in motion in a given volume is exactly equal to the number which come to rest in it, "electronic equilibrium" has been attained. This concept is illustrated in Figure A.2.

A.9 The Thimble Chamber. This type of measuring device is the most widely

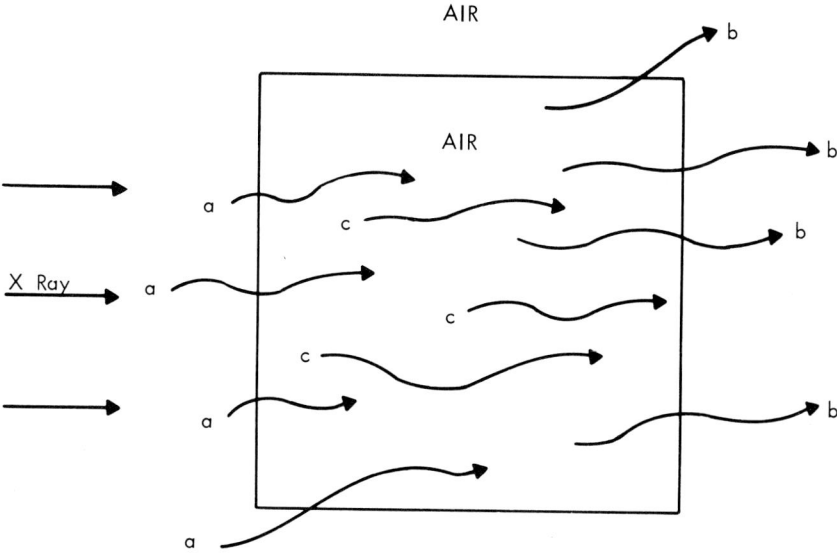

Figure A.2. A unit volume of air is shown, completely surrounded by air, and uniformly irradiated with x rays. There are three groups of secondary corpuscular radiations: (a) those with an origin outside the volume, but which lose a portion of their energy within it; (b) those which originate within the volume and lose a portion of their energy within it *but* produce ionizations outside the volume as well; and (c) those which originate *and* lose all their energy within the specified volume. When electronic equilibrium is attained, the loss from (b) is compensated by the gain from (a).

used instrument for the measurement of exposure (in roentgens). The large free-air chambers that are found usually in standardization laboratories are the primary standards against which the light and mobile thimble chambers may be calibrated at intervals. The air volume of thimble chambers is determined by the size and shape of the outer conducting wall which is similar in shape to the thimble used in sewing. The convenience of the thimble chamber relates to the replacement of the large thicknesses of air (large distances) required for electron equilibrium with lesser thicknesses of a more dense "air equivalent" with an effective atomic number equal to that of air. Thus, thimble chambers are built to artificially provide electron equilibrium. Consider the air volume in Figure A.3(a) as being surrounded by air whose pressure can be varied. If the pressure is doubled, secondary electrons can travel only half as far, and the air need extend only half as far in all directions. The

process could be "extended" so that the air is condensed to a "solid air wall." The status of the measuring volume itself will not have been changed, only the air surrounding it, as shown in Figure A.3(b). Thimble chambers have solid "air equivalent" walls made of a material that behaves in respect to x-ray absorption in the same manner as air.

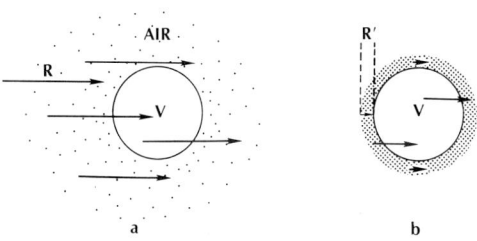

Figure A.3. Schematic diagrams of an air volume, V. a. V is surrounded by air in all directions to a distance greater than the range of a secondary electron, R. b. If the air is "compressed" down to a solid wall, the air will still extend in all directions to a distance equal to the *reduced* range of secondary electrons, R, in the more dense material.

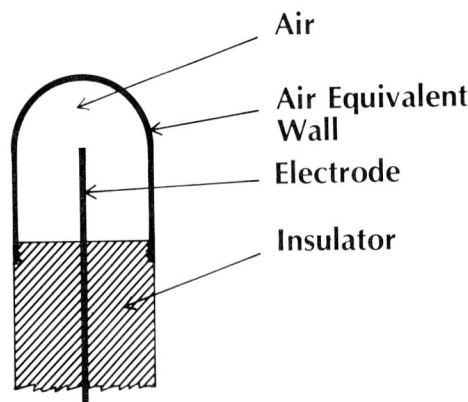

Figure A.4. Diagram of a thimble chamber with an air equivalent wall. The wall is usually Bakelite coated on the inside with carbon to make it electrically conducting.

A diagram of a thimble chamber with an air equivalent wall is shown in Figure A.4. The inside of the chamber is coated with a conducting layer, and an electrode is placed in the center of the air volume. If the electrode is charged, the ions producing the radiation will be collected. Thus, a given volume of air has been isolated (by the thimble volume) and the charge produced by the air ionizations has been collected. This fulfills the definition of the roentgen. The "air wall" of the thimble chamber must be as thick as the maximum range of the secondary electrons in that material. For x-ray voltages up to 70 KeV, thin walls such as those composed of nylon are used; for voltages from 70 KeV to several hundred kilovolts the wall is made of Bakelite of 0.5 mm thickness; for radiations greater than this (cobalt-60 at 1.25 MeV) a wall thickness of 3 to 4 mm must be provided.

A.10   The Mechanism of the Thimble Chamber.   The thimble chamber (also called a condenser chamber) uses a condenser system to store and measure electrical discharge produced by air ionization (Figure A.5).

The Victoreen condenser chamber represents an example of a widely used type of condenser thimble chamber. A diagram is shown in Figure A.6.

The thimble chamber with its attached condenser is inserted into the electrometer unit. The entire insulated system is then charged by an amber friction mechanism or by a transformer. The deflection of the system is detected by a movement of the platinum wire-quartz fiber assembly as viewed through an optical system and illuminated with a lamp from beneath. When fully charged, the hairline reads zero on the meter scale. The chamber is removed, a protective cap is placed over the connector end, and the chamber is placed within a radiation beam for a specific interval of time. The cap is then removed and the unit is reinserted into the fully charged electrometer. The loss of charge is indicated by the decrease in deflection of the fiber-wire assembly (the scale is calibrated in roentgens). The sensitivity of the instrument may be varied by using a variety of condenser thimble chambers and one electrometer.

Condenser chambers are secondary standards and should be calibrated against standard air chambers to determine the correction factor required to express the exposure dose in roentgens.

A.11   The Limitations of the Roentgen.   The definition of the roentgen (Section A.6) is restrictive in the following ways:

1. By definition it can apply only to x and gamma rays (the roentgen unit cannot be used for other radiations such as beta and alpha particles, neutrons, etc.).
2. All of the ionizations produced by the secondary electrons must be collected (electron equilibrium).

Thus, not only is the roentgen restricted to x or gamma rays, but it is also restricted to a certain energy range (specifically the roentgen is not defined for x or gamma rays with energies greater than 3 MeV). In order to satisfy the definition of the

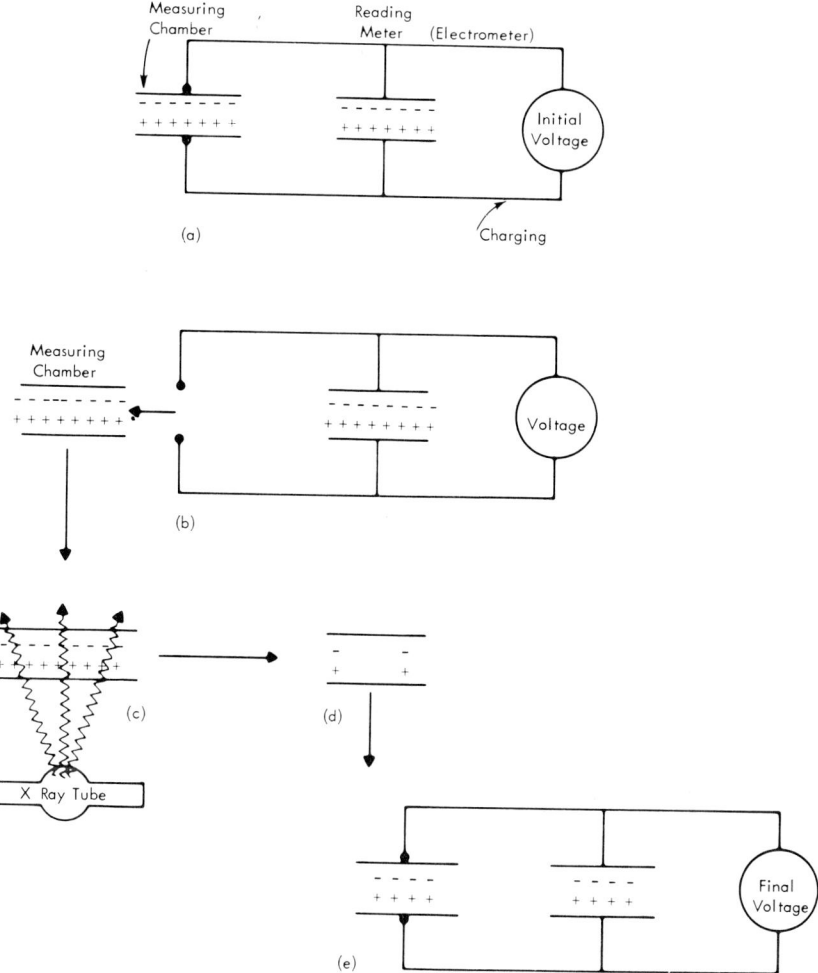

Figure A.5. A schematic diagram of a condenser chamber ionization instrument. (a) The charging mechanism charges the entire system, including the measuring condenser and the electrometer portion. The total voltage may be read. (b) The measuring chamber is removed, and (c) it is irradiated by an x-ray beam for a specified time period. (d) The radiation produces ionizations within the measuring chamber, and its charge in the chamber condenser and the electrometer condenser is shared, and a loss in charge (voltage) is measured. This voltage drop is proportional to the amount of radiation to which the chamber has been exposed.

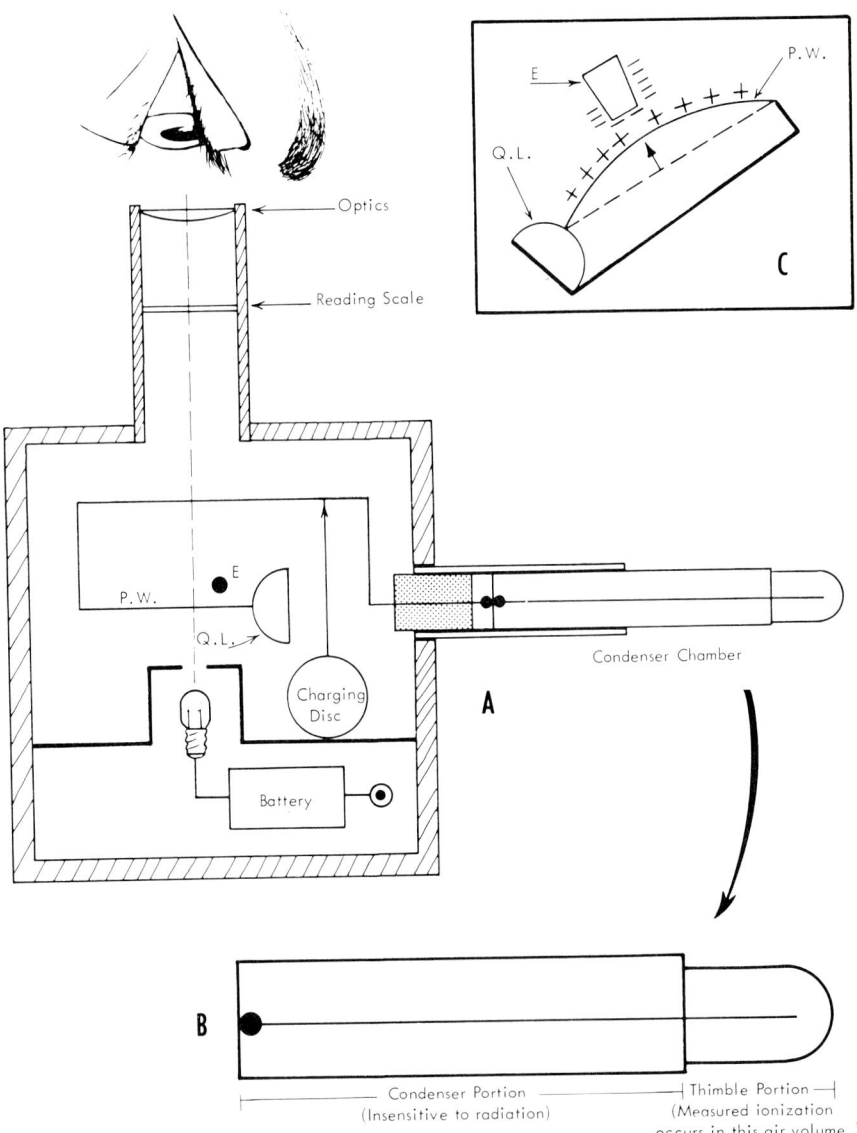

Figure A.6. Schematic diagram of a condenser r-meter (Victoreen). The condenser chamber with its thimble tip is shown: A, within the socket and electrically connected to the electrometer; B, in an enlarged representation removed from the electrometer. C. When fully charged, the platinum wire (P.W.) is deflected toward the deflection electrode (E) by movement against the quartz loop (Q.L.). The platinum wire forms the hairline image that is visualized on the scale. As the instrument is discharged by exposure to radiation, the platinum wire moves back toward its uncharged position. The deflection is proportional to the radiation dose.

roentgen, it is necessary for measurements to be carried out under conditions of equilibrium (in each successive layer of air traversed by the primary beam, equal numbers of secondary electrons must be generated and stopped). Scattered photons should also be avoided. As the primary radiation increases in energy, secondary electrons become more energetic (equilibrium is more difficult to obtain) and scattered radiations are unavoidable. At energies greater than 1 MeV, pair production with its energetic annihilation photons becomes important. Thus, for these higher energies, equilibrium conditions break down—the roentgen, as defined, cannot be accurately measured above 3 MeV.

Because of the restrictive nature of the roentgen, a new unit, the rad, was established in 1956.

A.12   The Rad.   The rad (a unit of *absorbed energy*) is the logical unit for radiobiology. That amount of energy which is actually transferred from a beam of radiation to the system being irradiated is the important portion, not the total energy to which the system is exposed. One rad represents 100 ergs per gram absorbed energy in any medium from any type of ionizing radiation. The rad is less restrictive than the roentgen (the energy absorbed from *any* ionizing radiation in *any* medium) and is a measure of the energy absorbed in a specified volume.

A.13   Measurement of Absorbed Dose: The Bragg-Gray Theory.   Direct measurement of energy absorbed within a given mass of biologic material is difficult. Biologic materials and the relatively low doses of radiation normally used in radiobiology do not lend themselves to techniques available for such measurements. However, the absorbed dose in a medium is directly related to the amount of ionization produced in that medium. If a small gas-filled chamber is placed within a medium, the ionization produced within it will provide an indirect measure of the energy being lost within the medium.

The Bragg-Gray concept forms the basis for the determination of absorbed dose from gas ionization measurements. The gas-filled chamber, however, must be small enough that its introduction into the medium does not affect the distribution or the numbers of secondary electrons in the medium.

An associated corpuscular emission (an electron) will not lose the same amount of energy in traversing a unit distance in the medium that it would in traversing the same distance in the gas. The relative mass stopping power (S) is a ratio of the ability of the medium to absorb the radiation per gram to the ability of the gas to absorb the radiation energy per gram—hence the concept of "stopping power."

Thus if S is the measure of the relative abilities of medium and gas to absorb the radiation energy on a unit mass basis, the energy absorbed per gram of the medium (E_m) can be calculated as:

$$E_m = S \cdot J \cdot W$$

This equation is the famous Bragg-Gray equation which allows one to calculate the absorbed dose in a medium (E_m) from the measurement of ionizations produced within a gas-filled chamber. If the charge (J) is measured in esu per unit volume of gas, W is known to equal approximately 34 eV/ion pair from gas ionization experiments, the only other quantity required is S (the mass stopping power ratio). The values of S are difficult to give precisely. If the medium and the gas have nearly the same atomic composition, S is nearly equal to 1 and does not vary much with energy. It is beyond the scope of this book to provide the needed tables of S values. Tables listing these values for various mediums, usually against air, are available in the literature.[2,3] These tables are generally limited in that the values listed are given in terms of the energy of the ejected electrons. Since x rays are emitted in a continuous spectrum, the ejected electrons will also have a spectrum of energies; therefore, average values are used to make this approximation.[3]

The utility of this approach is that the data may be used with any energy or any radiation. Although the roentgen has not been defined above 3 MeV, readings made with a small chamber of known composition (for instance, carbon) yield accurate absorbed dose values.

A.14 The Measurement of Absorbed Dose in a Medium with an Ionization Chamber Calibrated in Roentgens. A thimble chamber calibrated in roentgens may also be used to obtain absorbed dose in rads.

The medium and the air are exposed to the same radiation, but the fraction of the absorbed energy of the incident photons in each will be determined by the relative abilities of the air and the medium to absorb that energy. If the medium is also air, then an exposure of one roentgen will produce an absorbed dose of 0.87 rads in that air (in this instance the absorbed dose is less than the exposure dose). Tissues will absorb different amounts of energy from an exposure dose of one roentgen. The relative abilities of various media to absorb radiation (relative to air) have been determined, and are given as values of f. Thus f is defined as the factor which converts exposure in roentgens to absorbed dose in rads (the ratio of rads/roentgens) in a specific medium. Values of f have been calculated for a variety of different energies and are illustrated graphically in Figure A.7.

The conversion factors from roentgens to rads calculated for potential between 200 and 300 KV for the most common commercial x-ray machines are compiled by Goodwin[4] and offer a ready reference for this commonly used energy range.

An example will suffice to show the utility of this method.

A small calibrated dosimeter reads air roentgens correctly at half value layer of 0.5 mm copper. If the chamber is placed in a solid bone and an exposure dose of 100 roentgens is recorded, what will be the absorbed dose?
The f value for radiation HVL = 0.5 mm Cu is 2.6 (Figure A.7).
Thus the dose to the bone (D Bone) is:
D Bone = f × R rads
D Bone = 2.6 × 100 = 260 rads

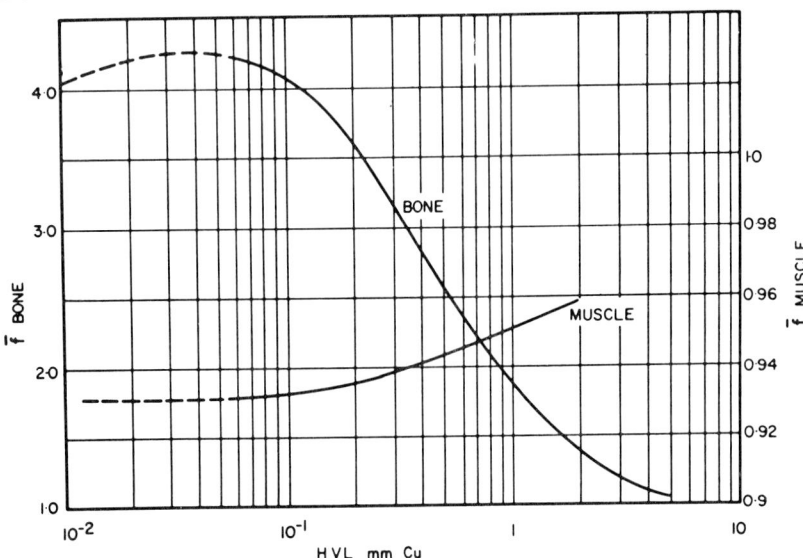

Figure A.7. Conversion factors f to convert from roentgens to rads for bone and muscle. (From International Commission on Radiological Units and Measurements [ICRU], 1956, Handbook 62. U.S. Nat. Bur. Standards.)

A.15 The Effect of Photon Energy on Absorbed Dose in Biologic Systems. It has been shown in the previous sections that absorbed dose in a medium may be obtained from measurements with an air- (or other gas) filled chamber. Indeed, in the photon range up to 3 MeV in which the roentgen is defined, a reading with a calibrated chamber may be converted to absorbed dose in the medium by use of the f factor (the ratio of roentgens/rads for the medium and air). This conversion factor is only slightly less than unity for the energy range of 10 KeV to 3 MeV for soft tissue while the conversion factor for bone varies from 4.5 to less than unity in the same energy range (Figure A.7). The conversion factor from roentgen to rads in air is a constant value (0.87), irrespective of the photon energy. This is because the roentgen is a unit of air ionization and the amount of energy required to produce an ion pair does not vary with the photon energy.

The fact that the relationship of absorbed dose to exposure (f) varies as it does is related to the composition of the various absorbing media: air, water (tissue), and bone. All these substances have, for practical purposes, the same number of electrons per gram. However, their effective atomic numbers are air = 7.64, water = 7.42, and bone = 13.8. At lower energies (less than approximately 60 KeV), the absorption of x- and gamma ray photons is primarily by the photoelectric effect which depends on the atomic number of the absorbing medium. Since water and air have essentially the same effective atomic number, they absorb radiation in about the same way. However, because of the higher effective atomic number of bone, the absorbed dose per gram of bone is much higher than air or tissue (Figure A.7). As the photon energy rises above these lower energies, the Compton process becomes more important. Compton scattering does not depend on atomic number but upon the number of electrons per gram of the absorbing medium (it is almost the same for air, muscle, and bone). Thus between 60 KeV and 200 KeV bone rapidly becomes less preferential in absorption. For photon energies between 200 KeV and 3 MeV the f factors for bone, water, and air are almost identical—there is no preferential absorption.

A.16 Energy Absorption within the Cavities of Bone. Bone consists of two kinds of material, a nonliving matrix of

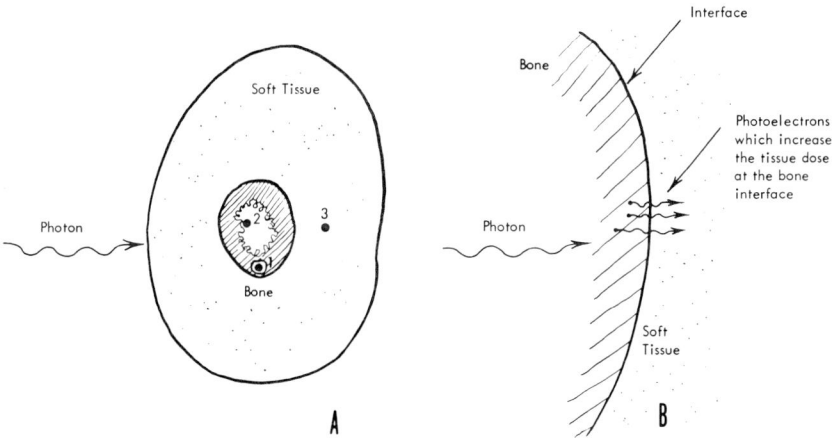

Figure A.8. Energy absorption within cavities of bone. A. The tissues of interest are 1, tissue within small cavities, 2, tissue within larger cavities, and 3, tissue shielded by bone. B. The dose to soft tissues falls rapidly as one moves from the bone-soft tissue interface because of the short range of the photoelectrons formed in bone. (Redrawn from International Commission on Radiological Units and Measurements, Handbook 85.)

calcium and within it the radiobiologi-cally important portions—the living cells associated with it. Bone may shield soft tissues that lie beyond it and decrease the absorbed dose to them, but soft tissues immediately adjacent to or enclosed within bone may have their dose *increased* by the contribution of secondary electrons arising primarily from photo-electrons formed from interactions with the calcium and phosphorus atoms of bone (Figure A.8).

If the tissue included within the com-pact bone has dimensions less than 1 μ, the dose received by the soft tissue will be the same as that for bone—the f factor may be applied directly. If, however, the dimension of the soft tissues is greater than 1 μ, the dose will be maximal at the bone-tissue interface, falling to a minimum at the center of the cavity (related to the range of photoelectrons formed in the bone as shown in Figure A.8).

The makeup of bone may be somewhat idealized to facilitate calculations of ab-sorbed dose (Figure A.9). In this idealized concept the osteocytes are found in small (5 μ) cavities throughout bone. Where the layers of bone are relatively thick, the osteocytes surround a haversian canal (50 to 100 μ diameter). These contain capil-laries, arterioles, veins, and arteries (as well as blood cells in transit through the vessels). The marrow spaces in trabecular bone (as opposed to compact bone), which contain the red marrow, are larger than the spaces of the haversian canals. They are on the average about 400 μ in diameter. In the haversian canals, os-teogenic cells form a lining on the bone approximately 10 microns thick. Table A.1 shows the difference in energy ab-sorption of these soft-tissue components of bone (this is given as the factor f, the conversion from roentgens to rads).

It is evident that there is a maximum radiation dose to soft tissues irradiated in

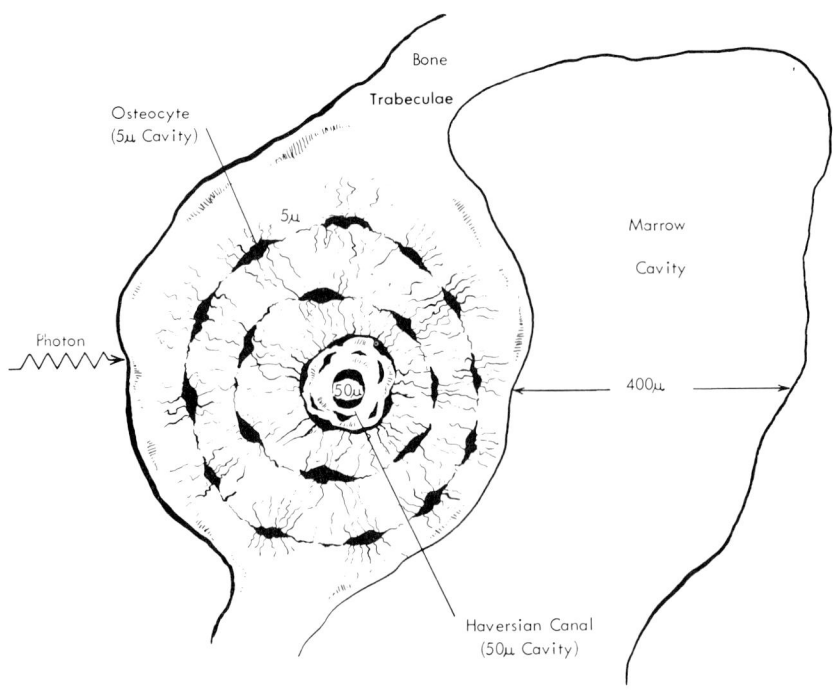

Figure A.9. Diagrammatic representation of the relationship of the soft-tissue components of bone. In regions of marrow the trabeculae are not usually large enough to have haversian canals, although one is shown in the diagram.

Table A.1. Mean Dose to Soft-Tissue Components of Bone*
(Mean dose factor f = rads/roentgen)

Photon Energy (KeV)	Soft Tissue (Muscle)	Compact Bone	Osteocyte 5 μ Diam.	10 μ Lining of Haversian Canal (50 μ)	Mean Marrow Dose for a 400 μ Marrow Space	Average Soft-Tissue Dose in Bone†
25	0.914	4.31	2.80	1.50	0.960	1.73
35	0.915	4.26	3.12	1.76	0.989	2.05
50	0.926	3.58	3.25	1.89	1.05	2.27
75	0.936	2.20	2.40	1.60	1.04	1.85
100	0.949	1.46	1.52	1.26	1.01	1.36
200	0.963	0.979	1.05	1.02	0.973	1.03

*This is an average value for osteocytes, cells in the haversian canals, and the bone marrow.
†ICRU Report NBS Handbooks 85 and 87 from data adapted from Spiers.

small bone cavities by photons of energies between 50 and 60 KeV. Further, the smaller the bone cavity, the higher the radiation dose to the tissue within it. The osteocytes within their 5 μ cavities are affected the most, the 10 μ cell lining of haversian canals to a lesser extent, and finally the red marrow cells of the bone marrow cavities (400 μ) the least.

SUMMARY

1. The amount of radiation directed at an object being irradiated is called the exposure dose. X- or gamma-ray exposure is measured in roentgens.
2. That portion of the radiation which is absorbed by the biologic system, the absorbed dose, is the biologically effective radiation. The unit of absorbed dose is the rad.
3. Absorbed dose may be obtained at a point within an irradiated medium by use of appropriate gas-filled chamber ionization measurements at that point.
4. At lower photon energies, high preferential energy absorption occurs within bone as compared to soft tissue.
5. At median photon energies there is no preferential bone absorption.
6. At higher photon energies, preferential bone absorption again occurs.
7. The living cells within bone are the radiobiologically important portions; their dose will depend upon the size of the cavity within the bone.

Text References

1. Bacq, Z. M., and Alexander, P.: Fundamentals of Radiology, 2nd Ed., Chapter 17, Delayed effects. New York, Pergamon Press, 1971.
2. International Commission on Radiological Units and Measurements. Handbook 85. Physical Aspects of Irradiation. U.S. National Bureau of Standards.
3. Johns, H. E.: The Physics of Radiology. Springfield, Ill., Charles C Thomas, 1961.
4. Goodwin, P. N.: Spectral measurements of 200 to 300 KVp x-rays. Radiology, 87:205$\sqrt{}$213, 1966.

Appendix B

IRRADIATION TECHNIQUES WITH EXTERNAL SOURCES OF X OR GAMMA RAYS

B.1 Introduction. The care in the determination of the radiation dose may vary from an estimate of a general level to a very precise measurement. An experiment should be designed so that the dosimetry is sufficiently accurate and reported in such a way that all pertinent information is given.

B.2 The Uniformity of Absorbed Dose. Usually in an experiment in radiobiology it is desirable to irradiate as uniformly as possible some well-defined tissue volume (in many instances this may be whole-body exposure). Even when only a portion of an organism is to be irradiated, uniformity of radiation dose is desirable, with as little radiation as possible given to surrounding tissue.

It has seemed useful to classify radiation conditions according to the degree of uniformity of absorbed dose.[1] These classifications are as follows:

  A. *Uniform irradiation* is defined as conditions under which variations in absorbed dose throughout the tissue volume are not large enough to have a significant effect on the biologic response. A ratio of less than 1.15 between the maximum and the minimum doses is considered "uniform irradiation."

  B. *Nonuniformity of irradiation due to absorption.* When radiations with limited penetrations are used to irradiate comparatively large animals, nonuniformity of absorbed dose is likely to occur. If the ratio between maximum and minimum dose does not exceed 1.30, this is called "moderately uniform." If, however, the ratio exceeds this value, the conditions are said to be "*nonuniform.*"

  C. *Nonuniformity in dose due to particle equilibrium.* When high-energy radiations such as cobalt-60 gamma rays are used, the dose rises rapidly and reaches a maximum at a depth of 3 to 5 mm in tissue (equilibrium depth). The tissues in the buildup region would be nonuniformly irradiated (see Class B). Thus the amount of incomplete dose due to unestablished equilibrium will depend on the thickness of the object and the energy of the radiation. For microorganisms exposed in air, considerable nonuniformity of dose may occur due to lack of equilibrium even with conventional x-ray energies.

It is usually desirable to have uniform irradiation (Class A). If this *is* attainable, one value will closely represent absorbed dose at any point within the tissue. If the dose is not uniform, it is the custom to express it as the "dose to the midline or the center" of the irradiated object or tissue volume.

B.3 The Effect of the Radiation Source.
When the results of irradiations of groups of specimens are to be compared, it is important that radiation output be closely reproducible. Modern x-ray machines are normally fairly well regulated with respect to tube current and potential (voltage). Small changes in tube current will produce proportionately small changes in output, but small changes in voltage may have a greater effect on the radiation output. Gamma-ray sources can be expected to be very constant except for decay; errors may occur if the source is partially shielded or they may be due to the presence of shorter lived radio-impurities.

When conventional x-ray machines are utilized to provide a large area of irradiation, the dose within this area may vary considerably from point to point. This type of nonuniformity may be averaged out by placing the objects to be irradiated upon a rotating turntable. However, the average dose should be determined by measurement with a dosimeter on the same rotating turntable.

B.4 The Effect of Distance from the Source.
It may be practically necessary at times to place an object (or animal) close to the source of radiation to obtain high dose-rates. If, however, the dimensions of the object are comparable to the source-object distance, considerable nonuniformity in absorbed dose will result because (1) the lateral extensions of the object will be further from the source

and receive less dose, and, (2) if the thickness of the object is significant, its proximal surface will be closer to the source than the distal surface. Tabulation of these variations may be found elsewhere.[1]

B.5 The Effect of Scattered Radiations.
X and gamma radiations may lose all their energy in an encounter with an orbital electron (photoelectric absorption) or undergo a scattering process in which their direction is changed. Radiations may be scattered into a tissue volume by material between the specimen and the radiation source, beside the specimen, or beneath the specimen (the latter is called backscatter).

These scattered and backscattered radiations can add materially to the absorbed dose in an object. The absorbed dose will rapidly rise as radiation scattering material (such as exposure supports, other animals, etc.) is brought close by. The addition of the *first* extraneous scattering material has the greatest effect on the absorbed dose. Additional scattering material contributes relatively *less* to absorbed dose. It usually facilitates irradiation to use conditions of *maximum* scatter (a condition in which the addition of further materials does not increase the absorbed dose). In addition, it facilitates the use of tables which give the reduction in dose with depth of absorber (depth-dose tables) and curves which show the distribution of dose in an absorber (isodose curves),[2,3] since these curves are determined with maximum scattering conditions. For photon irradiation it can be taken as a general rule that maximum scatter is obtained with substantial backing of materials whose density is equivalent to tissue (water and Masonite are good examples). The backing should be approximately 7.5 cm thick and exceed the width of the primary beam by 5 cm on all sides.[1] When irradiating groups of animals, there should be the *same* number of animals in each irradiated group each time the animals are irradiated. The animals should also be spaced uniformly. If equal groups of equal numbers are not to be used each time, *simulated* animals ("phantoms") should be added to make up the required number; dead or unwanted animals may also be used in this way.

B.6   Irradiation of Microorganisms. Although absorbed dose may be estimated and measured on the *macroscopic level* and clearly shown to be uniform, variations in absorbed dose may still occur if very small organisms are being irradiated (unestablished equilibrium during irradiation). Tissue culture cells which are attached to a glass surface may experience nonuniform irradiation because of photoelectrons ejected from glass which has a relatively high atomic number. Such variations are exceedingly difficult to detect and measure because they may extend only over a fraction of a millimeter. But they may seriously influence the outcome of a biologic study. Some of these problems may be overcome by placing such organisms within a container inside a large block of scattering material. When *low-energy* x-ray exposure is to be made of a thin layer of cells, the cells should be spread on something other than glass so as to reduce the effect of photoelectrons. Plastic dishes are available, but not all are useful; some may produce volatile, noxious compounds upon being irradiated. Plastics containing constituents of higher atomic number (polytetrafluoroethylene, vinyl chloride, etc.) should be avoided.[1]

B.7   Irradiation of Small Animals (Less than 250 Grams). When incident radiation is substantially absorbed within the volume being irradiated, nonuniformity of absorbed dose throughout the volume will occur. The portion of the irradiated volume *nearer* the source of radiation will absorb more radiation. This may be improved in two ways: (1) selecting a more penetrating radiation, and (2) irradiation by several sources at once or the same source from several directions (*multilateral* exposure). The most important step in the achievement of uniformity of absorbed dose is the progression from unilateral to *bilateral exposure.* Further extensions will achieve increased dose uniformity, but each of them will add only a small additional improvement. The kilovoltage required for moderate uniform irradiation of animals will depend on beam penetration and the size of the animal; however, the larger the animal the more necessary is bilateral irradiation.

Most therapeutic x-ray machines designed for clinical work are suitable for the irradiation of small animals. For x rays with a half-value layer greater than 1.5 mm of copper, the radiation is sufficiently

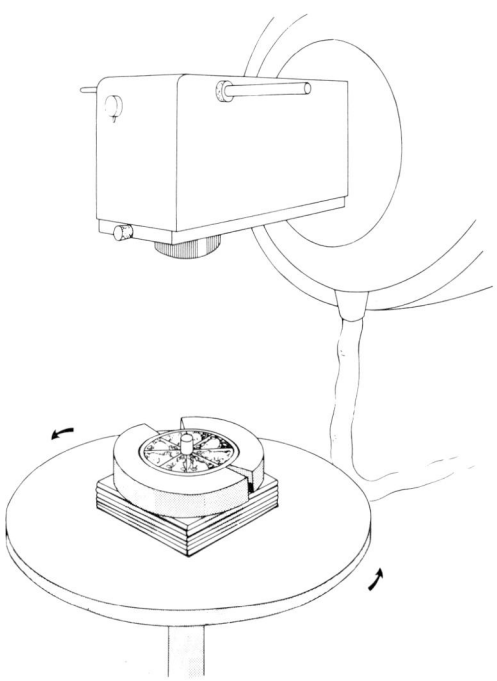

Figure B.1. Irradiation of small animals (mice) on a rotating turntable to improve dose uniformity. (Adapted from International Commission on Radiological Units and Measurements, Handbook 88.)

penetrating to ensure at least moderately uniform whole-body exposure of unilaterally irradiated animals as large as rats or medium-sized guinea pigs. However, for x-ray machines of less than 150 KV, bilateral exposure may be necessary for these animals to achieve even moderate uniformity of dose (unless the beam is highly filtered).[1] In any case, with x rays, uniformity of dose is improved by placing the animals in a Lucite irradiation cage and putting it on a rotating turntable (Figure B.1).

B.8 Irradiation of Medium-Sized Animals (250 Grams to 2.5 Kilograms). Medium-sized animals (rabbits, monkeys, large guinea pigs) usually require bilateral irradiation with x rays of at least 200 KV in order to obtain "moderate" uniformity of dose (Class B). Unilateral radiation may be acceptable for x-ray energies greater than 300 KV.[1] Medium-sized animals are best irradiated singly (in snug Lucite boxes) with adjacent material to yield maximum scatter conditions.

B.9 Irradiation of Larger Animals (Larger than 2.5 Kilograms). In the case of larger animals such as dogs and swine, bilateral exposure to x rays with energy greater than 250 KV may be adequate for moderate uniformity of dose, but bilateral exposure to supervoltage radiations or gamma rays is preferred.[1] Larger animals are best irradiated if they are "molded" so that the body is essentially circular with a minimum diameter. This is best accomplished by anesthetizing the animal and placing it in a circular container (Figure B.2).

B.10 The Recommended Exposure Technique and the Report of the Irradiation. The previous sections have described techniques that are useful in obtaining relative uniform absorbed doses in tissue volumes. In reporting the data of an experiment in radiobiology it is

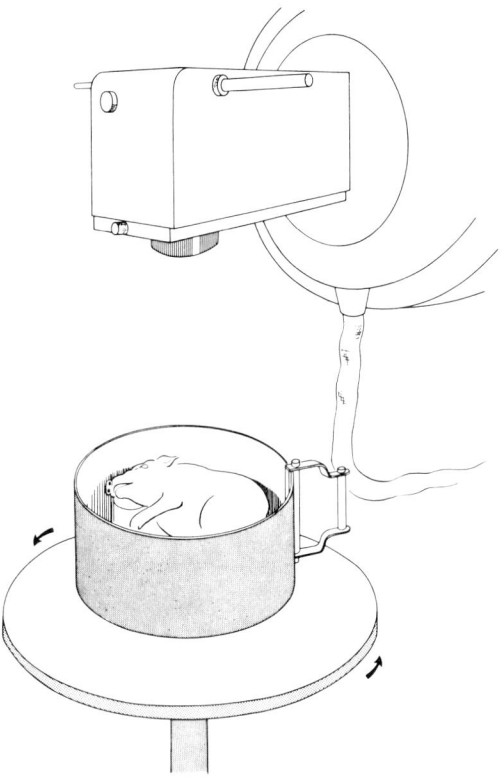

Figure B.2. Molding of the body of a large animal is illustrated. (Adapted from International Commission on Radiological Units and Measurements, Handbook 88.)

useful to describe the irradiation technique in complete detail so that the distribution of the absorbed dose and the technique are completely understood. An example of such a report is taken from ICRU Handbook 88 which describes a Class A (uniform) exposure of mice.[1]

"A constant potential x-ray machine was used to irradiate the mice, using the following exposure factors: 250 KV; added filtration of 0.5 mm copper, 1 mm aluminum; HVL, 1.2 mm copper; 30 ma; source distance (to center of animal): 100 cm. The mice were exposed, 10 at one time, in a circular container measuring 20 cm in diameter, divided into sectors, and placed on top of a block of wood measuring $25 \times 25 \times 7$ cm. The exposure with scatter was measured by placing a _____ (give make) dosimeter in the center of a phantom placed at a point corresponding to that of a representative animal, and the exposure rate thus determined, with the apparatus rotating, at approximately 3 revolutions per minute, was found to be 24 R per minute. The absorbed doses reported were derived by applying the factor 0.95. A diagram of the exposure arrangement used is shown in Figure _____."

This type of presentation is complete throughout and is an ideal model for other reports of animal exposure.

SUMMARY

1. Radiation is usually delivered to an organism or a tissue volume in such a way as to yield a relatively uniform distribution of absorbed dose. This uniformity may be influenced by:
   (a) The dimensions and thickness of the object with respect to the distance to the source,
   (b) Whether equilibrium conditions have been established, and
   (c) The effect of extraneous scattering materials.
2. Particularly for larger objects, bilateral irradiation, and irradiation with higher-energy x rays (or gamma rays), will significantly improve the homogeneity of the dose.
3. Examples of radiation techniques for small, medium-sized, and large animals are detailed with reference to recommendations by the ICRU.

Text References

1. Recommendations of the International Commission on Radiological Units and Measurements: Radiobiological Dosimetry. National Bureau of Standards, Handbook 88, Washington, D.C. (1963).
2. Johns, H. E.: The Physics of Radiology. Springfield, Ill., Charles C Thomas, 1961, pp. 704–744.
3. Glasser, O., et al.: Physical Foundations of Radiology. New York, Harper & Row, 1961, pp. 437–479.

INDEX

Page numbers in *italics* refer to illustrations;
page numbers followed by t refer to tables.

ABSORBED DOSE. *See* Dose, absorbed
Absorbing medium, 11
Absorption, energy, in bone cavities, 127–129, *127*
  nonuniformity of irradiation due to, 131
  photoelectric, 12, 13, 13t
  processes of, 11, 12–13
    preferential, 12, 13t
    primary, 13t
    relative importance of, 13–14, 13t
Accelerators, 10t, 11
Age, radiosensitivity and, 75
Air ionization, measurement by, 118–120, *119*
Air volume, compression of, 121, *121*
Alpha particles, 9, 10t
  bombardment of beryllium by, 11
  linear energy transfer values for, 16t
  tracks of, 17
Anemia, 104
  in hematopoietic syndrome, 79
Anencephaly, in rats, 95
Aneuploidy, 63
Animal(s), developmental abnormalities in, 94–95
  larger-sized, 134, *134*
  lethal dose for, 71–72, 71t
  life span of, 110–111, *111*
  medium-sized, 134
  small-sized, 133–134, *133*
Annihilation, 13
Anomalies, in mice, 95
Anophthalmia, in rats, 95
Antibiotic therapy, for radiation-produced infections, 104
Antibody(ies), cell-mediated, 99
  defined, 98
  humoral, 99
  production of, 99
    effect of radiation on, 100
    protection of, 100–101
    restoration of, 100–101
    small lymphocytes and, 99
    thymus and, 99–100

Antigen(s), defined, 98
Atom(s), electrons of, 8–9
  interactions between radiations and, 1, 9
  nucleus of, 8
    interactions between radiation and, 2
    stable configuration of, 9
  structure of, 8
Atomic number, pair production and, 13
  photoelectric absorption and, 12
Atomic reactors, 10t, 11

BACKSCATTER, 132
Bacteremia, in hematopoietic syndrome, 80
Bergonié and Tribondeau, law of, 68–69, 93
Beryllium, bombardment of, radiation produced by, 11
Beta particles, 10, 10t
Betatrons, 10, 14
Biologic damage, linear energy transfer and, 15
Biologic systems, absorbed dose in, effect of photon energy on, 127
  direct action in, relative importance of, 28–29
  indirect action in, relative importance of, 28–29
  radiation produced free radicals in, 27
  radiosensitivity of, hydration and, 26–27
Biology, radiation, defined, 1
Blood counts, in hematopoietic syndrome, 79
Body weight, radiosensitivity and, 75
Bone, composition of, 127–128, *128*
  conversion factor for, *126*, 127
  energy absorption in, 12, 13t, 128
  soft tissues of, mean dose to, 128, 129t
Bone cavities, energy absorption within, 127–129, *127*
Bone marrow, changes in, in gastrointestinal syndrome, 84, 86–87
  in hematopoietic syndrome, 77–79
  radiosensitivity of, 52–53
  replacement of. *See* Bone marrow transplants
Bone-marrow syndrome. *See* Hematopoietic syndrome

137

Bone-marrow transplants, autologous, 106
  heterologous, 101, 106
  homologous, 101, 106
  in irradiated individual, 101–102, 105–106
Bone tumors, 114–115
Bowel, function of, after irradiation, 69
Bragg-Gray theory, 125–126
Bragg peak, 16, 16
Brain, radiation damage to. See Cerebrovascular
      syndrome

CALORIMETRIC METHODS, for dose measurement,
      117–118
Cancer(s), 112–115
  induction of, steps in, 112, 115
  latent period in, 112
  radiation of, oxygen effect and, 48
  radiosensitivity of, 110, 112, 116
Carcinogen(s), 110, 112
Cataracts, radiation-induced, 115–116
Cell(s), analogy of, to solutions, 21–22
  changes in, in cerebrovascular syndrome, 90–91
      in gastrointestinal syndrome, 86
      in hematopoietic syndrome, 77–78, 79
      molecular, 27–28
  chemical reactions of, 1–2, 3
  chromosomal aberrations in, 65–66
  chromosome breakage in, 63–64
  compartments of, 67
  composition of, 21
  differentiating, 67, 69
  elimination of, after irradiation, 78
  extrapolation number for, 34
  function of, 2, 3
    loss of, 3–4, 5, 20
  functional, 67–68
  genetic mutations in, 58–66
  germ, chromosomes in, 62, 65–66
    radiosensitivity of, 109
  granule neuronal, after total-body irradiation, 90
  gut epithelium, renewal of, irradiation and, 85–86
  life cycle of, 41
    chromosomal aberrations and, 65
    genetic mutations and, 60
    radiosensitivity and, 41–43, 41, 42
    recovery and, 41–42, 42
  lymphoid, 100
  multitargeted, recovery of, 39–41
  oxygenation of, 44–45, 45, 47, 48
  precursor, gut epithelial, 86
  protective agents in, 6, 29
  radiations and, interactions between, 1, 4, 20, 28–29
  radiosensitivity of, 5–6, 35–38
    as function of oxygen tension, 44–45, 45
  recovery of, 4–5, 39–43, 41, 42
  reproductive capacity of, as survival end point,
      30–31
    loss of, 31, 32–33, 40
    radiosensitivity of, 35–36, 68
    survival of, 42, 42, 44
  somatic, chromosomes in, 62, 66
  stem, 67, 79
  survival of, 30–31, 31
    dose-rate and, 49–50, 50
    linear energy transfer and, 30, 48–49, 49

    oxygen and, 45–49, 47, 48
  synchronization in, 42–43
  targets in, 30–34
    in vivo, 54
  viability of, 3, 5, 60–61
Cell killing, 38
Cell kinetics, defined, 5
Central nervous system syndrome. See Cerebro-
      vascular syndrome
Cerebrovascular syndrome, 74, 89–92
  changes in, 92
    cytologic, 90–91
    histologic, 90
    inflammatory, 90
  death from, causes of, 91, 92
  manifestations of, 89–90, 91–92
  mean survival time for, 89, 91
  threshold for, 89
  vascular damage in, consequences of, 91
CFU. See Colony-forming units
Charged particle, energy dissipation, 14 15–16, 16
  interactions of, with matter, 14
  velocity of, 14–15
Chemical dosimetry, 118
Chemical reactions, cellular, 1–2, 3
  living matter and, 2–3
Chromosomal aberrations, cell cycle stage and, 65
  complex, 65
  dose and, 63–64, 65
  dose fractionation in, 65
  dose-rate and, 65
  linear energy transfer and, 64–65
  one-hit, 65
  significance of, 65–66
Chromosome(s), breakage of, cause of, 63
    consequences of, 63–64
  functions of, 62–63, 64
  homologous, 57
  loss of, 63
  number of, 62
Clotting time, in hematopoietic syndrome, 79
Colony-forming units, effect of radiation on, 52–53
Colosteoma, in mice, 95
Compton scattering, 12, 13t, 14, 127
Condenser chamber. See Thimble chamber
Crypt(s), intestinal, function of, 85
    in gastrointestinal syndrome, 83, 83, 84
    normal, 85
Crypt cell(s), radiosensitivity of, 86
    in mouse, 53–54
    regeneration of, 84–85, 85
Cyclotrons, 10–11, 10t

D₀, 36–37, 37
D₃₇, 36–37, 37
Death, causes of, in cerebrovascular syndrome,
      91, 92
    in gastrointestinal syndrome, 88
    in hematopoietic syndrome, 80
    pattern of, after total-body radiation, 74, 74
    species variations in, 71, 71t
Dehydration, treatment of, 105
Delta rays, 17, 17
Deoxyribonucleic acid. See DNA
Depth-dose tables, 132

Desiccation, radiosensitivity and, 26–27
Deuterons, 10–11, 11t
Dilution effect, 25–26, *26*
Distance, from radiation source, effect of, 132
Diurnal variation, radiosensitivity and, 75
DNA, *56*
  base sequences of, 57
  changes in, 3, 57, 58–59, *59*
  components of, 56–57
  functions of, 2, 57
  synthetic stage of, chromosomal aberrations
      and, 65
Dose, 17–18
  absorbed, defined, 129
    effect of photon energy on, 127
    measurement of, 125–126
    uniformity of, 131–132
  chromosomal aberrations and, 63–64, 65
  effects of, in total-body radiation, 71–73
  exposure, 129
  genetic mutations and, 59–60
  lethal, defined, 73
    for animals, 71–72, 71t
  life shortening as function of, 110–111, *111*
  mean, to soft tissues of bone, 128, 129t
  measurement of, 117–129
    air ionization, 118–120, *119*
    calorimetric, 117–118
    chemical, 118
    thermoluminescence, 118
  radiosensitivity and, 71–73
  relationship of, to leukemia, 113–114
  relative biologic effectiveness and, 18–19, *18*
  repeated, effect of, 112
  target inactivation as function of, 36, *37*
  threshold, 59
  zero, 36–37, *37*
Dose fractionation, chromosomal aberrations and, 65
  recovery and, 39–41, *39, 40, 41*
Dose-rate, chromosomal aberrations and, 65
  genetic mutations and, 60
  intracellular responses and, 49–50, *50*
  life shortening and, 111
  oxygen enhancement ratio and, 50
Dose-response curve, for crypt cells in mouse
      jejunum, 53
Dose-survival data, life cycle and, 41–42
Dose-survival response, exponential, 31, 32, *32*
  sigmoidal, 31–33, *32*
Dosimeters, chemical, 118
Dosimetry, chemical, criteria for, 118
Doubling dose, 62
Drug therapy, for radiation-produced infections, 104

EINSTEIN FORMULA, 13
Electrolyte(s), in gastrointestinal syndrome, 87
Electron(s), 8–9
  fast, energy distribution of, in water, 17
  ionization of, distribution of, 17
  kinetic energy of, 17
  linear energy transfer values for, 16t
  orbital, excited, 1, 9
    interactions between radiations and, 1, 9
    ionized, 1, 9
  rest mass of, 13
  tracks of, 17, *17*
  valence, 9
  velocity of, 15
Electron cloud, of atoms, 8–9
Electronic equilibrium, 120, *121*
Embryo(s), animal, developmental abnormalities in,
      94–95
  human, developmental abnormalities in, 95–96
  radiosensitivity of, 93–97
    clinical implications of, 96
  regeneration in, 93–94
Endocrine gland tumors, 114–115
Energy, absorbed, 18
  absorption of, 17–18
    within cavities of bone, 127–129, *127*
  alpha particle, 9
  average, per ionization, 9
  beta particle, 10
  conversion of, to mass, 13
  kinetic, 13, 17
    equation for, 14
    loss of, 15
  photon, 11
  transfer of, by excitation, 15
    by ionization, 15
    in complex systems, 20–21
Enzyme(s), changes in, implications of, 28
  dilution effect on, 26, *26*
Excitation, 9, 11, 15
  defined, 1, 19
Exposure, measurement of, 118–120
Extrapolation number, 33–36, *33*
  dose-rate and, 49, *50*
  for cells, *in vitro*, 54
    in vivo, 54–55
  oxygen and, 47, *47, 48, 48*
  radiosensitivity and, 35–36, *35*
  symbol for, 33

FEMALES, sterility in, 110
Ferrous sulfate dosimeter, 118
Fertility, defined, 110
  impairment of, 109–110, 116
Fluid(s), in gastrointestinal syndrome, 87
Free-air chambers, 120, *121*
Free radicals, 4, 22
  formation of, 22–23
  life of, 27
  radiation-produced, in biologic systems, 27
    sources of, 24
  reactions of, 23–24
    ion density and, 25–26
    oxygen and, 25
    products of, 24
    speed of, 23
  symbol for, 23
Fricke dosimeter, 118

GAMETES, chromosomes in, 62, 65–66
Gamma rays, 10, 10t. *See also* Photons
  external sources of, irradiation techniques with,
      131–135
  interaction of, with matter, 11
  uniformity of, 132

Gastrointestinal syndrome, 73, 82–88
  bone marrow in, 84, 86–87
  changes in, histologic, 84
  death from, causes of, 88
  electrolytes in, 87
  fluids in, 87
  infections in, 87–88
  intestines in, large, 85
    small, 84
  manifestations of, 82–84
  mitosis in, 86
  nuclear abnormalities in, 86
  precursor cells in, 86
  regeneration in, 84–86
Gene(s), complete set of, requirement for, 63
  DNA and, 56
  dominant, 58
  loss of, 64
  recessive, 58
  role of, 57–58
Genetic code, 57
Genetic equilibrium, 62
Genetic load, 61–62
Genetic mutations, 58
  cancer and, 115
  cell cycle and, 60
  dose and, 59–60
  dose-rate and, 60
  effect of, 60–61
  in large populations, 61–62
  lethal, 61
  point, 58–59, 59
  production of, 58–59, 59
    linear energy transfer and, 59
Genetics, radiation, 56–66
Genotype, 57
Gonads, radiosensitivity of, 109
Graft vs. host reaction, 101
Granulocyte(s), transfusion of, 104
Granulocyte count, in gastrointestinal syndrome, 87
Granulocytopenia, 103
  treatment of, 104

HEMATOPOIETIC SYNDROME, 73, 76–81
  blood counts in, 79
  changes in, cytologic, 77–78, 79
    histologic, 77
  death from, causes of, 80
  infection in, 79–80
  latent period of, 76, 78–79
  manifestations of, 76–77
  prodromal period of, 76
Hemorrhage, control of, 104–105
Hernia, in mice, 95
Heterozygous control, 57–58
Homozygous control, 57–58
Humans, embryo of, radiosensitivity of, 95–96
  life shortening in, 111–112
Hydration, radiosensitivity and, 25
    in living systems, 26–27
    in solutions, 25–26
Hydrocephalus, in rats, 95
Hypoxia, radioresistance and, 47
Hypoxic, 44

IMMUNE RESPONSE, enhancement of, 100
  primary, 98–99, 99
    radiosensitivity of, 100
  secondary, 98–99, 99
    radiosensitivity of, 100
  sequence of changes in, 99, 99
Immunity, acquired, 98–99
  alteration in, after radiation, 101
  innate, 98
  mechanism of, 98–99
  specificity of, 98
Immunoblast, 99, 99
Infections, in gastrointestinal syndrome, 87–88
  in hematopoietic syndrome, 79–80
  susceptibility to, after irradiation, 100, 101, 104
  treatment of, with drugs, 104
    with granulocytes, 104
Inflammation, in cerebrovascular syndrome, 90
Insertion, after chromosome breakage, 64
Intestine(s), large, in gastrointestinal syndrome, 85
  small, mucosa of, in gastrointestinal syndrome, 83,
      83, 84–85
  villi of, regions of, 85
Inversion, after chromosome breakage, 64
Ion clusters, 17
Ion density, free-radical interactions and, 24–25
  of charged particle, 15
Ion pair, 9
  formation of, 12
Ionization, 11, 12, 14
  defined, 1, 9, 19
  distribution of, from electrons, 17
  energy dissipated in, 9
  maximum, point of, 120
  specific, 15
Ionization chamber, calibrated in roentgens, 126
Ionizing radiation, acute, 62
  as carcinogen, latent period of, 112
  charged, 9, 10t
  chromosome breakage and, 63–64
  chronic, 62
  defined, 4
  direct action, 20–29
    defined, 4, 20, 21, 46
    protective agents against, 4, 29
    relative importance of, 28–29
  doubling dose for, 62
  effects of. See Radiation effects
  energy of, 9. See also Energy
  enhancement of, oxygenation and, 44, 46–47
  genetic mutation and, 58
  indirect action, 20–29
    defined, 4, 20, 21, 46
    mechanism of, 22
    relative importance of, 28–29
  interactions of, with atomic nuclei, 2
    with cells, 1, 4, 20, 28–29
    with matter, 8–19
    with orbital electrons, 1, 9
    with water, 22
  large populations and, 61–62
  linear energy transfer of. See Linear energy transfer
  partial body shielding from, antibody production
      and, 100–101, 101

quality of, 19
recovery from, 39–41
  cell life cycle and, 41–42
response to, dose and, 71–73
  in vitro, 51
  in vivo models of, 51–55
  pattern of, 72–73
scattered, 132–133
sources of, 9–11, 10t, 61
  effect of distance from, 132
split-dose, 39–41, 39, 40, 41
targets of, 3–4, 30–34
total-body, antibody production and, 100, 101
  effects of, 71–75. See also Cerebrovascular
    syndrome; Gastrointestinal syndrome;
    Hematopoietic syndrome; Radiation
    effects
  infection after, susceptibility to, 100, 101, 104
  lethality of, 7, 71–75, 74, 80
  mean survival time after, 72–73, 72
types of, 9–11, 10t
uncharged, 9, 10t
Irradiation, effect of. See Radiation effects; Radiation syndromes
immunity and, 100
manifestations of, 71, 76–77
  secondary, 103–104
nonuniform, due to absorption, 131
  due to particle equilibrium, 131–132
of animals, larger, 134, 134
  medium-sized, 134
  small, 133–134, 133
  techniques for, 133–134, 133, 134
of embryo, during organogenesis, 94–95
of microorganisms, 133
of solutions, 21
of water, 22
report of, 134–135
sterility after, 108–110
susceptibility to infection after, 100, 101, 104
techniques for, 131–135
uniform, 131
Isodose curves, 132

KIDNEY, radiosensitivity of, 69

LAW OF BERGONIÉ AND TRIBONDEAU, 68–69, 93
LET. See Linear energy transfer
Leukemia(s), 112–113
  granulocytic, 114
  in radiologists, 114
  lymphocytic, 114
  mortality rates from, 113
  mouse, response of, 51–52
  relationship of, to dose, 113–114
Leukemogenesis, radiation, in mice, 114
Life span, shortening of, 116
  dose and, 110–111, 111
  dose-rate and, 111
  in animals, 110–111, 111
  in humans, 111–112
Linear energy transfer, 15–17
  chromosomal aberrations and, 64–65
  defined, 15, 19

for ionizing particles, 15, 16t
genetic mutations and, 59
high, 30, 30, 49
intracellular responses to, 48–49, 49
low, 30, 30, 48, 49
oxygen enhancement ratio and, 44, 48–49
relative biologic effectiveness and, 18, 18
Lithium fluoride, for dose measurements, 118
Liver, function of, after irradiation, 69
Lucite irradiation cage, 133, 134
Lung cancer, 114
Lymphocyte(s), radiosensitivity of, 100
  small, antibody production and, 99
Lymphoid tissue, regeneration after, 77

MALES, sterility in, 109–110
Mass, conversion of, to energy, 13
Matter, interactions of, with ionizing radiation, 8–19
  with neutrons, 14
  living, chemical reactions and, 2–3
  particle tracks in, 15–17
  penetration of, by photons, 11
Maximum reparable damage, 54
Meiosis, chromosome breakage and, 64
Mesons, negative pi, 2, 10t, 11
Metabolic rate, radiosensitivity and, 75
Microcephaly, 95
Microorganisms, 133
Microphthalmia, in mice, 94–95
Mitosis, in gastrointestinal syndrome, 86
  in hematopoietic syndrome, 77–78
  radiosensitivity during, 41, 42
Molecule(s), changes in, chemical, 21
  significance of, 27–28
  irradiation of, 20
Mouse (Mice), abnormalities in, after irradiation
    of embryo, 94–95
  radiation leukemogenesis in, 114
  radiosensitivity studies in, on bone marrow, 52–53
    on crypt cells, 53–54
    on leukemia cells, 51–52
    on skin colonies, 54
Mouse jejunum, irradiation of, 53
Mucosa, of small intestine, degeneration of, 83,
    83, 84–85
Muscle, conversion factor for, 126
Mutagen(s), effectiveness of, 62
Mutant(s), 58
Mutations, genetic. See Genetic mutations

NEGATRONS, 10, 10t
Nervous system, damage to. See Cerebrovascular
    syndrome
Neutrons, 10t, 11
  fast, 14
  interaction of, with matter, 14
  linear energy transfer values for, 16t
  slow, 14
Normoxic, 44
Nucleotide(s), 57
Nucleus, atomic, 8
  interactions between radiations and, 2
  bone marrow cell, abnormalities in, 78, 86
  heavy, 10t, 11

OER. *See* Oxygen enhancement ratio
Organ(s), radiosensitivity of, 5–7, 67–70
    order of, 73–74, 75
    irradiated, secondary manifestations in, 103–104
        treatment of, 103–107
Organogenesis, embryo during, irradiation of, 94–95
Ovaries, irradiation of, 110
Oxygen, influence of, on free radicals, 25
    role of, in direct action, 46
        in indirect action, 46
    survival curves and, 47
Oxygen effect, 44
    mechanisms of, 45–46
    minimizing, in radiation of cancers, 48
Oxygen enhancement ratio, 46–47
    dose-rate and, 50
    linear energy transfer and, 44, 48–49
Oxygenation, 44
    degree of, 44–45, *45*
    intracellular responses to, 44–48, *45, 47, 48*
    mixed, survival curves of cells with, 47–48, *47*
    radiosensitivity and, 44–45, *45*, 48

Pair production, 12–13, 13t, 14
Particle charge, 14
Particle equilibrium, nonuniformity of irradiation
        due to, 131–132
Particle tracks, in matter, 15–17
Particle velocity, influence of, 14–15
Pauli's exclusion principle, 22
Phenotype, 57
Photoelectric absorption, 12, 13, 13t
Photons, 10
    absorption processes of, 11, 12–13
        preferential, 12, 13t
        primary, 13t
        relative importance of, 13–14, 13t
    high-energy, 12–13, 13t
    interactions of, probability of, 11
    low-energy, 12, 13, 13t
    median-energy, 12, 13, 13t, 14
    penetration ability of, 10
Photon energy, effect of, on absorbed dose, 127
Platelet(s), fresh, transfusion of, 105
    in hematopoietic syndrome, 79
Polyploidy, 63
Positrons, 10, 10t
Progression delay, 42, 43
Protective agents, cellular, 6, 29
Protons, 10–11, 11t
    linear energy transfer values for, 16t
    rest mass of, 15
    velocity of, 15
Pyknosis, of granule neuronal cells, 90

Quality factor, 19

Rad(s), conversion to, from roentgens, 126, *126*
    defined, 18, 19, 125
Radiation, ionizing. *See* Ionizing radiation
    sensitivity to. *See* Radiosensitivity
Radiation damage. *See* Radiation effect
Radiation effects, 1, 58. *See also* Radiation syndromes
    defined, 3, 5
    late somatic, 108–116

    modifying factors for, 108
    maximum reparable, 54
    on atomic nuclei, 2
    on cells, 1, 2, 3, 5, 27–28, 42
        chemical reactions and, 2–3
    on colony-forming units, 52–53
    on embryo, 93–97
    on fertility, 108–110
    on immunity, 98–102
    on life span, 110–111
    on orbital electrons, 1
    on organs, 5–7, 67–70
    on tissues, 5–7, 67–70
Radiation genetics, 56–66
Radiation leukemogenesis, in mice, 114
Radiation sensitivity. *See* Radiosensitivity
Radiation sickness, 108
Radiation syndromes, 73–74. *See also* Cerebrovascu-
        lar syndrome; Gastrointestinal syndrome;
        Hematopoietic syndrome
    terminal phase of, 79–80
Radiation tumorigenesis, 114–115
Radicals, free. *See* Free radicals
Radioactive decay, products of, 9, 10, 10t
Radiologists, leukemia in, 114
    United States, life spans of, 111–112
Radioresistance, *48. See also* Radiosensitivity
    of functional cells, 67–68
    of tissue, 6
    oxygen concentration and, 47
Radiosensitivity, 35–38. *See also* Radioresistance
    cell killing and, 38
    cell life cycle and, 41–43, *41, 42*
    curves of, as function of oxygen tension, *45*
    defined, 35, 36
    desiccation and, 26–27
    diurnal variation and, 75
    dose and, 71–73
    extrapolation numbers and, 35–36, *35*
    hydration and, 25–26
    indexes of, 36–37, *37*
    linear energy transfer and, 48–49
    metabolic rate and, 75
    of bone marrow, 52–53
    of cells, 5–6, 35–38
        differentiating, 69
        germ, 109
        lymphoid, 100
        multitargeted, 39–41
        precursor, 86
    of chromosomes, 64–65
    of embryo, 93–97
    of tissue, 6
    oxygenation and, 44–45, *45*, 48
Randomness, of particle interaction, 16
Rats, abnormalities in, after irradiation of embryo, 95
RBE. *See* Relative biologic effectiveness
Recovery, cellular, 4–5, 39–43, *41, 42*
    life cycle and, 41–43
Regeneration, bone marrow, 77, 78
    in embryo, 93–94
    lymphoid tissue, 77
    of crypt cells, in gastrointestinal syndrome, 84–85,
        85
Relative biologic effectiveness, defined, 18

dose and, 18–19, *18*
  expression of, 18
  linear energy transfer and, 18, *18*
Relative mass stopping power, 125
Repopulation, 6–7
Reproductive capacity, cellular, as survival end
    point, 30–31
  loss of, 31, 32–33, 40
  radiosensitivity of, 35–36, 68
  survival of, 42, *42*, 44
Restitution, after chromosome breakage, 63
Rest mass, of protons, 15
Roentgens, conversion of, to rads, 126, *126*
  defined, 18, 19, 120
  limitations of, 122, 125
  measurement of, 120
Runt disease. *See* Secondary disease

SECONDARY DISEASE, 101–102
  after bone-marrow transfusion, 106
Sex, radiosensitivity and, 75
Single hit:single target model, 31
Skeletal anomalies, in mice, 95
Skin cancers, 114
Skin colonies, mouse, radiation response of, 54
Solution(s), analogy of cell to, 21–22
  irradiation of, observations of, 21
  radiosensitivity of, hydration and, 25–26
Spermatazoa, mutation in, radiation induction of, 60
Spermatids, 109
Spermatocytes, 109
Spermatogonia, 109
Spermiogenesis, 109
Spina bifida occulta, in mice, 95
Spleen colonies, radiation response of, 52–53
Spontaneous mutation, 58
Sterility, 108–110
  complete, 109
  functional, 109
Stopping power, concept of, 125
Stress, radiosensitivity and, 75
Survival, dose and, 31
  dose-rate and, 49–50, *50*
  factors in, 44
  linear energy transfer and, 48–49
  of cells, 30, *31*
    end point of, 30–31
  pattern of, 74, *74*
  radiosensitivity and, 35–38
  recovery and, 39–43
Survival curves(s), dose-rate and, 49, *50*
  exponential, 30, 31, *31*
    semilogarithmic, 32, *32*
  for bone marrow, 53
  for cells of mixed oxygenation, 47–48, *47*
  for crypt cells, 53
  for mammalian cells, *42*
  for mouse skin colonies, 54
  in mouse leukemia model, 52
  in vitro, linear energy transfer and, 48–49, *49*
  oxygen and, 44–48, *47*
  sigmoidal, 30, 31–33, *31*
    components of, 31–32
    extrapolation of, 33, *33*
    semilogarithmic, 32, *32*
  split-dose, 39–40, *39*, *41*
    oscillation in, 42, 43

Survival time, for hematopoietic syndrome, 76, 80
  for gastrointestinal syndrome, 82
  for radiation syndromes, 73–74
  mean, 72–73, *72*, 83
    for cerebrovascular syndrome, 89, 91
Synchronization, in cells, 42–43
Synchrotrons, radiation produced by, 10–11

TARGET(S), 3–4, 30–34
  defined, 20
  inactivation of, curve for, 33–34, *33*
    dose and, 36, 37
    pattern of, 36–37
  number of, after split dose, 40, *40*
    estimation of, 33–34, *33*
    in vivo, 54–55
  radiosensitivity of, 35, *35*
  reappearance of, with time, *41*
Target molecules, 45–46
Target theory, 30
TD$_{50}$, for mouse leukemia, 51–52
Temperature, radiosensitivity and, 75
Testes, irradiation of, 110
Thermoluminescence, radiation-induced, 118
Thimble chamber(s), 120–122, *121*, *122*
  calibration of, 121, 122
  measurement of absorbed dose with, 126
  mechanism of, 122, *123*
Thrombocytopenia, 103–104
  treatment of, 104–105
Thymus, antibody production and, 99–100
Thyroid cancers, 114
Tissue(s), changes in, in cerebrovascular syndrome,
      90
  in gastrointestinal syndrome, 84
  in hematopoietic syndrome, 77
  lymphoid, 77
  radiosensitivity of, 5–7, 67–70
  soft, photon absorption by, 12, 13t
  of bone, mean dose to, 128, 129t
Translocation, after chromosome breakage, 64
Transplants, survival of, radiation and, 101
Tribondeau, Bergonié and, law of, 68–69, 93
Tumorigenesis, radiation, 114–115

UROGENITAL ANOMALIES, in mice, 95

VACCINATION, principle of, 99
Van de Graaf generators, 10–11, 10t
Vascular system, damage to. *See* Cerebrovascular
    syndrome
Vasculitis, in cerebrovascular syndrome, 90, 91
Viability, of cells, 3, 5
  mutations and, 60–61
Victoreen condenser chamber, 122, *124*
Villi, in intestines, *85*

WATER, cellular, 21
  interaction between radiation and, 22
  energy distribution of fast electron in, 17
  ionized, free radicals and, 22–23

X RAYS, 10, 10t. *See also* Photons
  external sources of, irradiation techniques with,
      131–135
  interaction of, with matter, 11
X-ray machines, radiation produced by, 10, 10t, 13
  uniformity of, 132